Predictability of Complex Dynamical Systems

Springer
Berlin
Heidelberg
New York
Barcelona
Budapest
Hong Kong
London
Milan
Paris
Santa Clara
Singapore
Tokyo

Springer Series in Synergetics

Editor: Hermann Haken

An ever increasing number of scientific disciplines deal with complex systems. These are systems that are composed of many parts which interact with one another in a more or less complicated manner. One of the most striking features of many such systems is their ability to spontaneously form spatial or temporal structures. A great variety of these structures are found, in both the inanimate and the living world. In the inanimate world of physics and chemistry, examples include the growth of crystals, coherent oscillations of laser light, and the spiral structures formed in fluids and chemical reactions. In biology we encounter the growth of plants and animals (morphogenesis) and the evolution of species. In medicine we observe, for instance, the electromagnetic activity of the brain with its pronounced spatio-temporal structures. Psychology deals with characteristic features of human behavior ranging from simple pattern recognition tasks to complex patterns of social behavior. Examples from sociology include the formation of public opinion and cooperation or competition between social groups.

In recent decades, it has become increasingly evident that all these seemingly quite different kinds of structure formation have a number of important features in common. The task of studying analogies as well as differences between structure formation in these different fields has proved to be an ambitious but highly rewarding endeavor. The Springer Series in Synergetics provides a forum for interdisciplinary research and discussions on this fascinating new scientific challenge. It deals with both experimental and theoretical aspects. The scientific community and the interested layman are becoming ever more conscious of concepts such as self-organization, instabilities, deterministic chaos, nonlinearity, dynamical systems, stochastic processes, and complexity. All of these concepts are facets of a field that tackles complex systems, namely synergetics. Students, research workers, university teachers, and interested laymen can find the details and latest developments in the Springer Series in Synergetics, which publishes textbooks, monographs and, occasionally, proceedings. As witnessed by the previously published volumes, this series has always been at the forefront of modern research in the above mentioned fields. It includes textbooks on all aspects of this rapidly growing field, books which provide a sound basis for the study of complex systems.

A selection of volumes in the Springer Series in Synergetics:

Yurii A. Kravtsov
James B. Kadtke (Eds.)

Predictability of Complex Dynamical Systems

With 55 Figures

 Springer

Professor Dr. Yurii A. Kravtsov

Space Research Institute, Russian Academy of Sciences,
Profsoyuznaya 84/32, 117810 Moscow, Russia

Dr. James B. Kadtke

Institute for Pure and Applied Physical Sciences,
University of California at San Diego,
9500 Gilman Drive, La Jolla, CA 92093, USA

Series Editor:

Professor Dr. Dr. h.c.mult. Hermann Haken

Institut für Theoretische Physik und Synergetik der Universität Stuttgart
D-70550 Stuttgart, Germany
and
Center for Complex Systems, Florida Atlantic University
Boca Raton, FL 33431, USA

```
Library of Congress Cataloging-in-Publication Data
Predictability of complex dynamical systems / Yurii A. Kravtsov, James
B. Kadtke (eds.).
       p.    cm. -- (Springer series in synergetics, ISSN 0172-7389 ;
vol. 69)
    Includes bibliographical references and index.

    1. Prediction theory.   2. Differentiable dynamical systems.
3. Nonlinear theories.   I. Kravtsov, IUrii Aleksandrovich.
II. Kadtke, James B., 1957-    . III. Series: Springer series in
synergetics ; v. 69.
QA279.2.P73   1996
519.5'4--dc20
                                                      96-31944
```

ISBN-13 : 978-3-642-80256-0   e-ISBN-13 : 978-3-642-80254-6
DOI : 10.1007 / 978-3-642-80254-6

Typesetting: Camera-ready by editors
SPIN 10426020   55/3144 - 5 4 3 2 1 0 - Printed on acid-free paper

Dedicated to the Kadtke and Kravtsov families,
who have always proven quite *un-predictable*...

Preface

This book was originally conceived as a continuation in theme of the collective monograph Limits of Predictability (Yu.A. Kravtsov, Ed., Springer Series in Synergetics, Vol. 60, Springer-Verlag, Heidelberg, 1993). The main thrust of that book was to examine the various effects and factors (system non-stationarity, measurement noise, predictive model accuracy, and so on) that may limit, in a fundamental fashion, our ability to mathematically predict physical and man-made phenomena and events. Particularly interesting was the diversity of fields from which the papers and examples were drawn, including climatology, physics, biophysics, cybernetics, synergetics, sociology, and ethnogenesis. Twelve prominent Russian scientists, and one American (Prof. A.J. Lichtman) discussed their philosophical and scientific standpoints on the problem of the limits of predictability in their various fields. During the preparation of that book, the editor (Yu.A.K) had the great pleasure of interacting with world-renowned Russian scientists such as oceanologist A.S. Monin, geophysicist V.I. Keilis-Borok, sociologist I.V. Bestuzhev-Lada, historian L.N. Gumilev, to name a few. Dr. Angela M. Lahee, managing editor of the Synergetics Series at Springer, was enormously helpful in the publishing of that book.

In 1992, Prof. H. Haken along with Dr. Lahee kindly supported the idea of publishing a second volume on the theme of nonlinear system predictability, this time with a more international flavor. It was then that the present editors (Yu.A.K. and J.B.K.) agreed to produce all the materials. During the ensuing period of preparation, Dr. Lahee happily gave birth to a baby (an event which could hardly have been predicted by either of the editors before the project began; we only wish for Angela that her child will prove a little more predictable than any of ours), and Prof. W. Beiglböck subsequently took over management of the project. Prof. Beiglböck wisely advised us to change the emphasis of the second volume, to provide a more practical and technical approach to the theme. Hence, this second volume provides the reader with a more applied framework, which is oriented towards the analysis of experimental data or artificial signals, often providing explicit numerical algorithms. The resulting book is both interesting and unique, in that the papers provide philosophical, intuitive, analytical, and numerical aspects of these state-of-the-art methods of analysis, which are drawn from experts in a wide variety of fields. The intent is, of course, cross-fertilization between these diverse disciplines, and we hope the reader will find them just as highly interesting and instructive as we have.

We would like to express our great appreciation to all the authors of this monograph for the timely submission of their papers and for their patience during the rather extended period of preparation. We are especially thankful to Dr. M. Kremliovsky for his heroic efforts in the TEX formatting of the camera-ready copy of the manuscript. We are also indebted to Ms. E.B. Grigor'eva for the superb translation of three Russian papers into English. It is our sincere wish that the joint efforts of all the participants of this project has resulted in an interesting, readable, and technically useful book.

Moscow   *Yu.A. Kravtsov*
San Diego   *J.B. Kadtke*
June 1996

Table of Contents

List of Contributors

Oleg L. Anosov
Vladimir Regional
Cardiology Center,
Sudogorodskii Rd. 67
Vladimir, 600023, Russia

Jeffrey S. Brush
Cambridge Research Associates
1430 Spring Hill Rd, Suite 200
McLean, VA 22102, USA

Oleg Ya. Butkovskii
Vladimir Technical University
Gorkii Str. 87
Vladimir, 600029, Russia

Robert Cawley
Information Sciences and
Systems Branch
Naval Surface Warfare Center
Dahlgren Division, White Oak
10901 New Hampshire Avenue
Silver Spring, MD 20903-5640, USA

Guan–Hsong Hsu
Information Sciences and
Systems Branch
Naval Surface Warfare Center
Dahlgren Division, White Oak
10901 New Hampshire Avenue
Silver Spring, MD 20903-5640, USA

Kevin Judd
Centre for Applied Dynamics
and Optimization
University of Western Australia
Nedlands 6907, Western Australia,
Australia

James B. Kadtke
Institute for Pure and Applied
Physical Sciences
University of California, San Diego
9500 Gilman Dr., La Jolla
CA 92093, USA

Holger Kantz
Department of Physics
University of Wuppertal
Gaußstrasse 20
D-42097 Wuppertal, Germany

Yurii A. Kravtsov
Space Research Institute,
Russian Academy of Sciences
Profsoyuznaya 84/32
117810 Moscow, Russia

Michael N. Kremliovsky
Institute for Pure and Applied
Physical Sciences
University of California, San Diego
9500 Gilman Dr., La Jolla
CA 92037, USA

Gottfried Mayer-Kress
Center for Complex Systems
Research, Beckman Institute and
Department of Physics
University of Illinois at
Urbana-Champaign
Urbana, IL 61801, USA

Alistair I. Mees
Centre for Applied Dynamics and
Optimization
University of Western Australia
Nedlands 6907, Western Australia,
Australia

Nikita N. Moiseev
Computational Center of Sciences,
Vavilov Str. 40,
117333 Moscow, Russia

Martin Paulus
Laboratory of Biological Dynamics
and Theoretical Medicine
Department of Psychiatry
University of California, San Diego
La Jolla, CA 92093, USA

James B. Ramsey
Department of Economics
New York University
New York, NY 10003, USA

Liming W. Salvino
Information Sciences and
Systems Branch
Naval Surface Warfare Center
Dahlgren Division, White Oak
10901 New Hampshire Avenue
Silver Spring, MD 20903-5640, USA

Thomas Schreiber
Department of Physics
University of Wuppertal
Gaußstrasse 20
D-42097 Wuppertal, Germany

Zhifeng Zhang
Department of Mathematics
Stanford University
Stanford, CA 94305, USA

Part 1

Introduction

Introduction

James B. Kadtke and Yurii A. Kravtsov

1 A Changing Paradigm

As recently as 50 years ago, there was a firm conviction among many scientists that the universe was fundamentally mechanistic, and that at some
level mathematical prediction of physical events could be exact. This view
was of course rooted in the "romantic" period of the history of science (the
19th and beginning of the 20th centuries) when the overwhelming advances
in science and technology often obscured the possibility that fundamental
limitations to the power of science could exist. The situation has changed
rather dramatically in recent decades. Most scientists will agree now (as did
the most acute minds long ago) that long-term mathematical prediction of
complicated physical systems is in practice unachievable. What is remarkable is that this realization has evolved from the "precise" science of classical
mechanics, which has long upheld the principles of Laplacian determinism.
This is due, ironically, to the development in the last twenty years of a consistent framework for "chaotic" dynamical systems, often generalized now to
"nonlinear dynamics". These ideas have required an essential revision of the
concepts of dynamical behavior and classical predictability.

Because of these advances, there is now an enormously renewed interest
in classical dynamical systems and modeling, and in particular time series
analysis and prediction. Much of this work attempts to take advantage of the
rather unusual properties of systems exhibiting chaotic behavior. Typically,
such a system exhibits so much local instability that any small inaccuracy
in the specification of the state variables is rapidly magnified (sometimes
referred to as the "butterfly effect"), typically making long-term predictions
impossible. Since this type of generic instability is typically produced by nonlinear terms in the evolution equations of a continuous physical system, the
study of this field has taken on the term nonlinear dynamics; as the classes of
systems (and types of behavior) studied have become more general, we now
speak in terms of the field of complex systems. Another important property
is that, in many instances, the existence of chaos in a dynamical system can
mean that a far simpler model of any complicated time evolution can be developed to explain its behaviour (due to the existence of a low-dimensional
"attractor"), and this latter possibility has resulted in an explosion of numerical algorithms being developed to take advantage of this possibility. To
evaluate and understand the performance of these methods, new and objective criteria need to be developed that are tailored to the dynamical modeling
framework. For example, a useful concept introduced by Lighthill [Lighthill,

1986] is the "horizon of predictability", which is the farthest future time beyond which no prediction based on dynamical models can make sense. Not only does chaos present a new understanding of where this horizon may lie, but in many instances can point out the difference between an objective scientific prediction, and a subjective one based on dubious means. At present, the "new dynamicists", in many fields, are busily struggling with these problems, and it was the original intention of this monograph to examine some of these issues.

The theme of this current monograph may at first seem somewhat obscure, and indeed required a good deal of thinking to organize in an easily digestible form. One must realize, though, that modeling is a fundamental aspect of science (and other fields), where the practitioner builds a mathematical framework premised on some fundamental hypotheses, and then uses this to compare a model-based prediction to reality. The fact that our conceptualization of what constitutes modeling and prediction is changing within the framework of complex systems means that some fundamental aspect of modern science may also change. It is not the purpose of this book to delineate what these changes may be, since these sciences are currently evolving rapidly, and we are certainly not expert enough to tell. However, we here do attempt to bring together a variety of new ideas from some of the most important experts in their respective fields, to present technical examples and methods which examine some aspect of the predictability issues which may not be readily available in the literature. More importantly, we have attempted to collect such papers from a wide range of fields, with the intention of making available to the interested professional some of the common problems and solutions which are faced in disparate disciplines. It is the intent of the book to promote such "cross-fertilization" between these disciplines, which often face fundamentally identical technical problems, yet develop fundamentally different solutions.

A second hopefully interesting aspect of this book is that the editors have tried to encourage an intuitive, even speculative flavor in the contributions, and not simply technical discussions. This is in major part because a hallmark of the new area of complex systems is a fundamentally changed view of how computers can be used for research purposes. In the last two decades, the vastly increased power of computers, and new languages and visualization methods, have been coupled with the many new theories and modeling procedures to allow the researcher to explore qualitative aspects of the system of interest. In this sense, the researcher has become an "experimenter" in the virtual laboratory of his computer, and intuition and imagination have become key aspects. Consequently, modeling and prediction have taken on considerably more importance, even in the physical sciences, as researchers now search for global properties, rather than simply examining strings of numbers. Such strategies are perhaps not as foreign in, e.g., the social and political sciences, where modeling has often been the primary means of analysis. For example, model-based prediction may be used to explore the physically feasible behaviors existing in the universe of all possible behaviors, or for simple contingency planning. The central point here is that such intuitive analysis is becoming a key part of the research process even in the physical sciences, and will likely become more so in the future. We have therefore viewed it to be vital that the contributions in this volume contained at least some discussion of the qualitative or speculative aspects of each individual problem, and we feel that largely this has proven much more enlightening. It is this intuitive aspect that has provided the development of whole new paradigms for understanding many natural phenomena in recent years, and will no doubt continue to do so.

Perhaps a bit more should be said about the importance of prediction and predictability in the understanding of complicated physical systems. As hinted at above, the new sciences developing in complexity theory utilize at least two fundamentally new realizations: first, that a system whose time evolution may seem exceedingly complex (even to the extent that standard mathematical measures may declare it to be "random") may in actuality have simple nonlinear, deterministic models which can generate its behavior. This first concept has led in recent years to the development of whole new paradigms in some fields (e.g. cellular automata, or "artificial life"), as scientists attempt to re-interpret old, "random" data in light of new deterministic models. The second concept is that even simple deterministic systems can exhibit such local instability (i.e., chaos) that any mathematical prediction loses validity after some short average time scale. Although the implications are less well understood at this stage, this second point may have profound consequences for many fields of study which are model-oriented, especially in the social, military, and political arenas. Understanding when systems are predictable, and especially when they are fundamentally not, can lead us away from erroneous questions and lines of reason, and toward more relevant understanding. For example, in simulating a socio-military-political conflict, generals or politicians can move towards the physical parameters of the system which result in regular, predictable behavior, in order to avoid dynamical regimes which can generate catastrophic "flashpoints" which cannot be foreseen or controlled. As such, nonlinear modeling may have profound consequences for many aspects of society, and new ways of quantifying and understanding the predictability of given systems will be a vital aspect. It is hoped that many of the papers of this volume have at least touched on the issues involved in this aspect, and provide some examples of the technical framework available to researchers. We should also point out that many of these philosophical issues are also discussed in the excellent collective monograph *Long-Term Predictability* (Shebehely and Taply, Eds., 1976, Dordrecht: Reidel) and also in the discussion stimulated by J.Lighthill in "Proceedings of Royal Society", **A407** (1832), 1986.

The present book contains 12 contributions collected from an international group of researchers, of which 8 are from the US, 4 from Russia, 3 from Germany, one from the UK, one from Australia, and one from China. The papers are organized into six chapters, which progress from the most fundamental issues associated with predictability and data analysis (i.e., Chap. 2: The Search for Determinism; and Chap. 3: Modeling and Forecasting Algorithms), and then move on to specific fields and applications (Chap. 4: Prediction of Biological Systems; Chap. 5: Analysis and Forecasting of Financial Data; and Chap. 6: Socio-Political and Global Problems). Although the papers are fairly technical, this progression is such that inspection of the first few papers will rapidly provide the novice researcher with sufficient background to understand the concepts discussed in the latter application papers.

2 Is It Deterministic?

In Chap. 2, four papers are presented which address perhaps the most fundamental issue in nonlinear dynamical data analysis and modeling. That is, determining when a set of data (that may appear random to linear techniques) may actually possess deterministic structure, caused by nonlinear correlations in the data, and hence is amenable to representation by a nonlinear dynamical system. Each of the four papers deals either directly or indirectly with this issue, and these discussions nicely set the stage for the remainder of the book. The first paper, "Method to Discriminate Against Determinism in Time Series Data", by Cawley, Hsu, and Salvino, provides an intuitive and entertaining discussion of the generic steps a researcher might go through when confronted with a complex data set, if he wishes to take advantage of dynamical modeling techniques. Afterward, the authors attempt to define, from a practical standpoint, exactly what "determinism" means in an observed time series, and how it can be identified. Here, the authors make the case that determinism can be identified directly with a property they call "smoothness". It is perhaps not easy to explain in a simple way what smoothness is, but we may simply say that determinism in a system implies a smooth evolution of state space points, which may be considered differentiable. This concept of determinism may differ slightly from other conventions used in data processing, in particular from the concept of "partial determinateness", which equates determinism with complete predictability, and randomness with perfect unpredictability, with a gradual transition existing from one extreme to the other [Kravtsov and Etkin, 1981, 1984; Kravtsov, 1989, 1993]. Nevertheless, the convention of treating determinism as smoothness is useful for many applications since it allows a clear procedure for discrimination between deterministic chaos, with underlying smooth phase trajectories, from noisy data which are less smooth. The authors develop a quantitative measure of smoothness (W, the "index of smoothness") which is easy to calculate. Using this quantifier, one may try to distingush between "smooth" dynamical objects with $W = 1$ and noisy ones with $W \ll 1$. The authors also demonstrate the usefulness of the smoothness index by some rather impressive numerical examples. It is certain that not all dynamical systems belong to the class of smooth ones, but in fact the concept presented here embraces the essential class of objects of practical interest for physical systems. In summary, the first step in any analysis is understanding when a set of data may imply that the underlying generator is deterministic, and hence is amenable to a dynamical modeling scheme, and the authors have presented a new and practical definition for this which should be considered closely.

The second paper of Chap. 2 is entitled "Observing and Predicting Chaotic Signals: Is 2% Noise Too Much?", presented by Schreiber and Kantz. This deals with another fundamental issue which is particularly important for researchers dealing with real data. To understand this, consider the following scenario: one may record a time series of data from a system which one may suspect or know to be a deterministic dynamical system, yet the data is corrupted by small random components (i.e. "noise" from measurement errors, external sources, etc.), which is of course almost unavoidable in practice. The question is now whether even a small level of such noise destroys our ability to model and predict the deterministic components, or worse still, calculate dynamical measures, or even to classify the signal as deterministic. If the answer to this question is generically "yes", then in its present form nonlinear dynamical modeling would have a grim future in the world of real applications. In this paper, however, the authors present some rather thorough and well-conceived discussions which address the effects of noise on prediction times, and also the calculation of dynamical invariants such as Lyapunov spectrum and entropies, which are often taken as indicators of determinism. In general, the conclusions here can be roughly summarized by stating that most dynamical measures of determinism (e.g. correlation dimension) are reasonably robust to small amounts of noise, but as the noise level approaches a few percent, estimates can become quite unreliable. Often, however, it is possible to derive simple scaling relations which can at least mitigate some of the false conclusions which may arise because of this. While these results imply that caution is necessary, they do not preclude the use of the dynamical approaches for a typical data set. As far as predictability issues are concerned, the authors show that the situation can often be much worse, sometimes because the process of fitting a model can itself become unreliable at these noise levels. In fairness, the authors here consider a relatively restricted class of models for their results, and do not consider more sophisticated approaches which may have superior noise mitigating properties. In general, though, this paper provides clear and intuitive explanations of the issues involved in any such analysis, and also provides some simple rules-of-thumb for the scaling properties with noise of many of the dynamical measures. In all, this paper is a very useful contribution.

It should be noted that it has been previously shown from a different point of view [Lighthill, 1986; Kravtsov, 1989, 1993] that the time of predictable behavior depends directly on the intensity of the noise level, derived from whatever corrupting physical processes are present. A rough estimate for the predictability time for chaotic systems has the form

$$\tau_{\text{pred}} = \frac{1}{2\lambda_+} \ln(A/\delta) \, ,$$

where λ_+ is the largest positive Lyapunov exponent, A is the approximate attractor size, and δ characterizes the noise level in the system. The logarithmic dependence of τ_{pred} on the "signal-to-noise ratio", A/δ, enters into this equation due to the exponential divergence of trajectories in locally unstable systems. This weak, logarithmic dependence of τ_{pred} on the noise level is of great importance. It implies that every real (chaotic) physical system may be predicted only for a comparatively short time interval, which has been aptly named the "horizon of predictability" [Lighthill, 1986]. It is the horizon of predictability which is rather sensitive to the noise level, and a 2% relative noise level may indeed be sufficient for a strong degradation of predictability.

The third paper in this chapter is entitled "A Discriminant Procedure for the Solution of Inverse Problems for Non-stationary Systems", contributed by Anosov and Butkovskii. Here, the authors present some very simple but highly instructive examples of the power of nonlinear modeling techniques in the analysis of time series of data. The particular setting is that of discriminant analysis — that is, determining when two or more segments of data (or different signals, usually measured close together in time) may have slightly different properties, indicating a change in parameters of the physical generator, or a variety of other effects. Such problems are of prime importance in, e.g., signal encoding, automatic machine control, etc. The particular data examples the authors use are simulated data sets constructed by evolving chaotic maps or differential equations, which are additionally made *non-stationary* by changing the parameters of the underlying system either smoothly or abruptly during the course of the experiment. A main point here is that standard discriminant analyses, which are typically based on comparison of statistical measures such as mean and variance, are almost completely insensitive to changes in these data sets, since the chaotic signals appear to be largely "noise" to these linear measures.

The authors then demonstrate some rather remarkable results by using an alternate procedure: they first fit a simple nonlinear dynamical model to the data sets, in two sliding, time-consecutive windows. From these estimated models they construct a discriminant function (here essentially the difference in the estimated parameters), and also construct a statistical estimator from these based on a Fisher criterion. Using this algorithm, the authors then show that rather small changes in parameters (less than 0.5%) can be easily detected for either smoothly or abruptly varying changes in the generating functions. Although the authors do not discuss the effects of additive noise or model mis-specification, results by other authors indicate that such approaches are fairly robust to these effects, in the sense that the discriminant performance scales smoothly and in a straightforward fashion with their magnitudes.

The importance and essence of these results is that it is easy to construct examples of data sets for which *the correlations in the data are primarily nonlinear*, and hence standard linear analysis methods will be insensitive to subtle (or sometimes large) changes in the system parameters. The use of nonlinear modeling methods can often reveal this structure, and thus can prove quite sensitive to "hidden" characteristics of the data. Although this basic concept may itself seem quite simple, the current development of such analysis techniques are a result of the vast increases in computer power in recent years, and also in the recognition and deeper theoretical understanding of the dynamics of highly unstable dynamical systems. As these methods become more well understood and widespread, there is no doubt that they will begin to have a very significant effect on many fields requiring processing of complex data types.

This general analysis scheme is taken to a more sophisticated level in the final paper of this chapter, "Classifying Complex, Deterministic Signals", by Kadtke and Kremliovsky. Here, the authors consider the more general problem of being able to distinguish between signals of unknown origin or generating form, which may also be highly corrupted by external noise components. This problem can be specified more exactly as either a detection problem (determining if and when a deterministic signal exists in an otherwise long stretch of random components), or a classification problem (placing an observed signal into one of possibly many categories of signals that have been observed previously). Detection and classification are fundamental areas of signal and data processing, and have wide importance in sonar, radar, and image processing, as well as in physics, geophysics, voice recognition, and biomedical applications (e.g., EEG analysis to classify physiological states) to name a few.

In their discussion, the authors consider only sample data sets which are constructed to have strong nonlinear correlations (not necessarily chaotic), and scale away the linear signal information by windowed renormalization, since the interest lies in determining the improvement in performance beyond the linear techniques. They then delineate how an entire detection/classification processing chain can be constructed algorithmically, where a key aspect is the fitting of nonlinear dynamical models (sets of ordinary differential equations, or "global models") to short data segments. The central and innovative aspect here, however, is that the authors have redefined their analysis procedure as an *observation* problem; that is, the relevant information lies in the statistics of the feature space of many short data observations, represented through the estimated coefficients of the "dynamical filter". By taking this approach, it is possible to directly assign classification probabilities and performance criteria, in analogy with standard detection theory. Using these algorithms, the authors demonstrate some remarkable detection and classification performance on a variety of very complicated signal types, including data drawn from open-ocean acoustic recordings of whales and dolphins. An additional advantage of this method is also its generality, since no *a priori* assumptions are made about the model class used, or even the stationarity of the signal and underlying model class. Using these techniques, the authors present numerical examples which can in some cases detect and classify different signal observations that are corrupted by as much as 300% noise.

As in the third paper of this chapter, this contribution is meant to show that utilization of nonlinear correlation information (not necessarily for chaotic behavior) can often yield significant improvements in modeling and analysis methods, particularly for data which appears too "complex" or noisy for standard linear analyses. In the next chapter, we consider methods which make explicit use of dynamical properties of the models in question (including chaotic behavior), and specifically consider the problem of forecasting.

3 Exploiting the Dynamics

In Chap. 3 we continue the discussion of some fundamental issues associated with nonlinear dynamical methods for data analysis, but here we move on to the more sophisticated issues of modeling and prediction. This chapter contains three papers by several experts in this area, dealing with both the modeling procedure itself as well as some of the implications of these ideas. An essential concept to grasp in this chapter is that, for data sets which contain components generated by dynamical systems, the optimal predictor must in theory be a model which exactly captures the detailed dynamical structure of the underlying generator, since this represents the maximum amount of information which is obtainable from the system. Optimal extraction and utilization of this information, particularly in the presence of noise, then presents a technical issue, requiring clever numerical methods. Hence, the modeling and prediction issues are intimately intertwined for dynamical techniques. The papers in this chapter discuss both issues, and provide some excellent insights and examples of their application.

The first paper in Chap. 3, "Strategy and Algorithms for Dynamical Forecasting", is presented by Anosov, Butkovskii, and Kravtsov, and provides a good general discussion which sets the stage for the remainder of the chapter. In it, the authors discuss the general framework for dynamical modeling, which they term the "inverse problem of dynamical systems", and discuss its application to prediction or forecasting. The particular approach discussed is that which attempts to estimate a set of ordinary differential equations (i.e., a dynamical system) which may govern the physical system, either in actuality, or as an empirical model for the effective motion of the observed dynamics. The authors point out that such algorithms for dynamical reconstruction were implied many years ago, most succinctly by A.N. Kolmogorov and O. Gabor, who considered the solution of such nonlinear inverse problems. Methods for the modeling of *chaotic* dynamical systems are quite analagous, with the only fundamental difference being the short predictability time of the chaotic systems due to their local instability. It is easy to show that for systems with nonlinear correlation, the dynamical predictors can be no worse than (and are typically much better than) predictors derived from statistical approaches, which are limited by the autocorrelation time τ_{corr}; similarly, standard predictors developed from linear (autoregressive) approaches are also limited by τ_{corr}, and nonlinear autoregressive methods do only slightly better. Since nonlinear correlations can often act over much longer time scales than the linear correlations, inclusion of such information is highly desirable. Hence, if a data set does indeed possess significant nonlinear correlations (perhaps discovered by the methods of Chap. 2), then such nonlinear predictors can take advantage of this additional information, and can often perform far better than standard predictors.

The numerical method that the authors outline is fairly sophisticated, and uses a number of new ideas which attempt to reduce the effects of a variety of numerical problems. For example, the processing chain consists of initial pre-filtering, estimation of the system dimensionality, construction of an optimal basis in the Takens time-delayed framework, transformation to a new basis which includes first-derivatives of the time series, and finally elimination of unreliable coefficients using an instability criterion. As anyone who has ever attempted to solve an inverse problem well understands, such sophisticated methods are often necessary simply to ensure numerical stability. Here, the authors have apparently developed a rather stable and reliable scheme, as they present several numerical examples where impressive results are obtained. Although the authors due not strictly measure the performance of their extracted models as predictors, or compare them to linear methods, the fact that the original dynamical system can be almost exactly recovered for some of the multi-dimensional data sets indicates that near-optimal use is made of the information available.

It should be pointed out that a variety of techniques for reconstruction of underlying dynamical systems have been published in recent years [see, e.g. Crutchfield and McNamara, 1987; Cremers and Huebler, 1987; Kadtke and Brush, 1991; Baake et al., 1992], which are generally termed "global modeling" methods. Many of these use quite sophisticated numerical schemes, and have in many cases yielded remarkable results. In addition, an entirely different approach using only local dynamics ("local linear" methods) on an attractor has also been developed extensively [see, e.g. Farmer and Sidorowicz, 1987], and provided very impressive results. The interested reader is encouraged to explore this growing volume of literature to obtain a flavor for these powerful methods.

The second paper of Chap. 3 is entitled "Parsimony in Dynamical Modeling", and is contributed by Mees and Judd. Here, we find another example of a dynamical modeling method for observed, possibly noisy data, however of quite a different mathematical form. The authors concentrate on the class of pseudo-linear functional forms (i.e., linear combinations of nonlinear functions), and their particular choice is the well-known *radial basis* functions [e.g., Casdagli, 1989]. This approach points out the seeming dichotomy which exists in the modeling methods used for dynamical systems: researchers can typically choose either a global, *functional* representation of the data evolution, such as the empirical sets of differential equations described in the previous paper, and which may have no specific concern about the underlying phase space geometry of the particular data set. The second class of methods typically used are *geometric* in nature; that is, they explicitly reconstruct a phase space for the data evolution, examine the geometric structure explored by the data, then build what often amounts to an extrapolation scheme to generate local (usually one-step) predictions in the phase space [e.g. the "local-linear" method of Farmer and Sidorowicz, 1987]. This second class is somewhat more unique to dynamical systems theory, since an important concept in the field is the direct correspondence between dynamical and geometric structure. Since the latter may have no simple mathematical representation, these methods can sometimes be purely geometric (i.e., algorithmic) and quite general, although they usually must inherently assume the existence of an attractor for the particular data set. The radial basis function approach which the authors present here is an interesting intermediate, since they explicitly utilize the specific geometric structure of the attractor as in the local-linear methods; however, they construct a predictive model using relatively few analytic functional forms valid throughout the entire data set, and hence are simultaneously somewhat global in nature.

Although the authors discussion of radial basis functions is quite enlightening, their properties are not the main point of the paper. Rather, Mees and Judd use this framework to introduce the important concept of "parsimony" in data modeling. Simply put, consider the well-known idea that any data set can be modeled exactly, if one is allowed to construct a large and complicated enough model (i.e., with sufficiently many paramaters). In so doing, however, the resulting model must capture all the real or artificial nuances of the data of interest, including the effects of inherent (random) noise. If one's aim is to contruct a dynamical model for a deterministic component of the data, then this situation is of course fundamentally contradictory. Hence, faces the task of constructing the *simplest* model, which still captures the *essential* dynamics of the data. This basic concept of parsimony is fundamental to many fields where modeling is applied, and is essential in data applications where random components are present; however it is only recently being introduced in dynamical systems modeling.

Even though parsimony is a succinct and intuitive concept, it is often extremely difficult to understand and enforce in a practical application. In this paper, the authors present a discussion and several good examples of how this may be done algorithmically. They present a simple, iterative "pruning" algorithm for the radial basis function formulation, which reduces unnecessary parameters in the model. They also present an excellent discussion of a rigorous approach to this problem, based on the work of Rissanen and others. This approach equates the parsimony problem with the encoding problem, and considers the best model to be the most efficient code which reproduces the data set. This idea is summarized quantitatively by the Minimum Description Length (MDL) principle, which has received much attention in recent years. Finally, the authors present several numerical examples which nicely demonstrate some of their concepts on simulated and real data. For those readers who are currently using dynamical modeling methods, this paper is worth close study, since the use of parsimonious algorithms is beginning to prove essential in current research applications. Alternate approaches and further discussion of these ideas as applied to dynamical modeling can be found in other references such as [Crutchfield and McNamara, 1987; Gribkov et al., 1994; Kadtke and Brush, 1994] and recent works by Sidorowicz.

The final paper of Chap. 3 is entitled "The Bifurcation Paradox: The Final State Is Predictable if the Transition Is Fast Enough", by Butkovskii, Brush, and Kravtsov. Here, we move away from describing the particular methods used for nonlinear dynamical modeling, and concentrate on an example of how new properties of a physical system may be discovered by extracting and examining a dynamical model. The general concept to note here is that, if one is capable of constructing a reliable and valid model of a particular physical system, from either one or an ensemble of data observations, then the researcher has the possibility of inferring new (unobserved) properties of the system by investigating the behavior of the model itself. Such approaches are particularly important for the rich behavior exhibited by strongly nonlinear dynamical systems, since e.g. multiple attractors may co-exist in a system; however, a researcher may only be capable of observing a small sample of the possible behaviors inherent in the dynamics.

The particular setting that this paper discusses concerns the behavior of a dynamical map model which may be derived from some physical data by parameter estimation. One of the most fundamental ways of understanding the global characteristics of such dynamical systems is to outline the "bifurcation" structure of the system. Simply put, this means that we chart out the types of solutions which the system generates as various parameters of the model are changed. For example, for some ranges of parameters the system may generate simple oscillatory behavior of a particular frequency. However, as the parameters are changed (through a "bifurcation point") this behavior may suddenly result in oscillations with e.g. twice (or half) the frequency. Studying the qualitative character of the system's solutions over all available parameter ranges gives a complete, global picture of the system's behavior.

Within this setting, the authors ask the following question: if we examine a dynamical system for which the system parameters are smoothly changing (non-stationary), and we can measure or model the system behavior *before* a bifurcation point, can we then make any prediction about the state of the system immediately *after* it has passed through the bifurcation point? Surprisingly, the authors show that the state of the system after bifurcation can indeed be predicted, even in the case of noise corruption. The key to this predictability lies in the speed of the parameter transition, which must be sufficiently large compared to the (bandwidth limited) noise influences. Intuitively, the reasoning here is that under fast transition the role of noise becomes insignificant, because the noise dynamic simply has not sufficient time to equalize the probability of the two possible final states of the bifurcated system. If the transition is not fast enough, however, then the conventional equality in probabilities of final states is dominant, and no exact prediction is possible. The authors demonstrate this interesting phenomenon with a number of convincing numerical examples.

A physical framework for understanding this result may be that of symmetry breaking in the nonlinear system. Such broken symmetries are quite important in many physical (e.g., laser polarization) and biological (e.g., the origin of life) phenomena. The effect mentioned above may highlight the problem of violated symmetry from a fresh point of view. Interestingly, this phenomenon may have important implications for those attempting to model the transitions in socio-economic-political systems. The authors point out that a similar situation is occurring in the former Soviet Union and East Bloc countries, as they attempt to move (smoothly) from a centralized to a free-market economy, and from a communist to a democratic political system. In this case, prediction of the final state through a bifurcation point could prove vital, as they attempt to avoid economic instabilities or political chaos.

4 Chaos in Biology

Chapter 4 of our book moves us away from the theoretical and fundamental aspects of dynamical modeling and forecasting, and begins our discussions of applications to particular fields. One of the most promising and challenging fields for new applications of dynamical systems and complexity is biomedicine, which is the subject of the two papers in this chapter. Because of the sheer complexity, synergism, and variability of biological systems, they have eluded close mathematical description for generations of scientists, yet the new analysis and modeling methods of nonlinear dynamics are already having some success in these areas. The reader is generally referred to work by Haken, Mackey, Glass, Rapp, and others for some important insights.

The first paper of this chapter is entitled "Models and Predictability of Biological Systems", contributed by Paulus. This paper provides an excellent and concise introduction to the basic concepts which must be considered when attempting to model or predict biological systems, and contains many relevant references and examples. The author begins by trying to define what prediction means in the context of biological systems, then goes on to discuss how one understands what to predict, the purposes of prediction, how one characterizes a predictive model, the statistical framework necessary to account for inherent errors, and the limits of predictability for any biological system. The author also provides two examples of practical numerical techniques which can quantify structure in observed noisy and complex biological data, from which can be inferred important properties of the system under study. The author nicely points out that prediction for biological systems is a fundamental medical tool, and gives examples of applications such as assessing survival rates of pathological conditions like cancer, or prediction of physiological states (such as sleep staging) from EEG records, or modeling human physiology to determine time-varying dosages of medication.

One of the main points that should be drawn from this paper is that biological systems can be extremely difficult to model because they typically operate on widely disparate time and length scales simultaneously. Hence, the system may display structure on some or many of these "spatio-temporal" scales, and appear random on many others. Therefore, the principal issues the researcher needs to ask are a) what does he wish to model, and how can he utilize it; and b) is there sufficient structure in the time series of the observed variables to build a useful model to accomplish his task? For the second point, Paulus describes a numerical tool which may be quite important for choosing the time and length scales which exhibit the most structure. This numerical scheme, called Correlation Integral Difference Surfaces (CIDS), is relatively easy to compute, and may well be worth utilizing as a pre-analysis tool before modeling of a system is attempted [see Paulus et al., 1993].

To summarize, this paper shows that far more can perhaps be done in the mathematical characterization and prediction of biological systems than was imagined even ten years ago. On the whole, this paper is highly recommended for those readers who would like a brief overview and practical guide to this rapidly developing field.

The second paper of Chap. 4 is entitled "Limits of Predictability for Biospheric Processes", contributed by the distinguished Russian Academician Moiseev, and is rather quite different in tone from the rest of the papers of this monograph, being almost philosophical in nature. Here, the author asks the rather daunting question of what predictability means in the context of the physical sciences as a whole, and as applied to Mankind's understanding of the physical universe. The author begins with an almost historical perspective of the modern scientists' conceptual framework of the Universe, then immediately asks the question of how it may ever be possible to understand systems of such immense scope and complexity using modern scientific tools.

To provide some hint to these questions, Moiseev examines a specific example of the modeling of a highly complex system, by addressing the problem of the stability of the Earth's biosphere. Put more simply, he asks whether the complex interaction of atmosphere, oceans, plant and animal life (and humans), etc. is stable to relatively minor perturbations of, e.g. carbon dioxide content, or whether a global ecological disaster can be initiated with these changes. This question has of course been the subject of hot scientific and political debate in recent decades, and whether or not we can hope to reliably model such a phenomenon may of course be vital to Mankinds future. Interestingly, via simple models and mathematical stability analysis, Moiseev shows that rather important qualitative characteristics of such model systems can be simply derived. The author then examines a hierachy of models for the biosphere, each including greater sophistication, and considers the stability (or, rather, the "sustainability") of each of them in turn. One point demonstrated is that while a given model may produce regular, stable solutions, inclusion of only slightly more sophistication in the model can produce quite different (even chaotic) dynamics. Therefore, the author argues that while individual models can often provide significant insights, the principal problem is usually understanding when sufficient sophistication (e.g., number of relevant variables) have been built into the model, so that in a sense the heirarchy itself is consistent.

Although at the end of the paper the author can make no definite statements about the stability of our biosphere, the work provides an excellent example of how simple mathematical modeling and analysis can provide important *qualitative* insights, even if the system under study may be too complex to ever fully characterize.

5 Complexity in the Financial Markets

In the fifth chapter of this monograph, we move to an entirely different field of study: economics. Although somewhat apart from the physical sciences, the study of economic systems is itself evolving rapidly, due to the advent of computerization as an analysis tool and a practical trading mechanism. The study of economics also has vast importance for a variety of reasons: firstly, the mathematical and statistical study of financial markets has absorbed enormous resources in the West for at least a century, both by pure economists and investors, since these markets are of immense importance to the growth of industrialized countries, as well as involving huge amounts of capital. Secondly, these systems are among the most complex known to man, and have always eluded a close mathematical analysis. For example, it is still not known whether the dynamics of particular indicators of the US economy are largely random or deterministic, or whether they are internally or externally "driven". The new tools of nonlinear dynamics may help to answer some of these questions in the future. Thirdly, forecasting and estimation of the limits of predictability are obviously of primary importance to the investors and economic planners in our society, and understanding qualitative aspects of the dynamics itself are quite valuable. One example of this is the changing effect on the US economy as we move toward a global economy. Many argue that, even if the economy is now approximately a low-dimensional dynamical system, its dimension will certainly rise as we are coupled to the world economy. However, others have drawn the opposite conclusion, saying that the explosion in the use of automated computer-trading algorithms may be imposing structure on the entire system, and hence actually lowering the dimensionality.

As an example of the new analysis methods being developed in the economic community at present, we have included a very interesting paper by Ramsey and Zhang, entitled "The Application of Wave Form Dictionaries to Stock Market Index Data". The paper is rather technical in nature, and does not deal directly with prediction or predictability; however, it provides an excellent overview of the relevant issues in this area. The authors present a new algorithmic method for characterizing and analyzing time series of complex data, specifically using the Standard and Poor 500 economic index of prices as an example. Although this method is not strictly dynamical in nature, it is one of many new numerical methods falling under the category of spatio-temporal and scale decompositions, such as "wavelet" analysis. The general concept here is that exceedingly complicated (perhaps random-appearing) signals can be decomposed and visualized into a spectrum of scales, locations, and waveforms, much like a Fourier decomposition represents stationary signals as a spectrum of single frequencies. Such methods can be exceedingly useful in providing intuitive understanding of the hidden structure inherent in such signals, since complicated signals are often composed of a few simple waveform constituents which are related in a nonlinear fashion, or may be obscured by random components. Decomposing these constituents can often aid in constructing simple models which capture the qualitative features of the behavior.

In this paper, the authors state a number of important results. They first point out that financial time series such as the S&P 500 are typically *not* random, but the structure which is exhibited is generally weak. More importantly, the data are fundamentally non-stationary, and hence require modeling and analysis tools which account for such non-stationary dynamics. Along these lines they then discuss an impressive generalization of wavelet analysis, called "waveform dictionaries", and provide some rigorous analytic results for their properties. They also indicate how this tool is to be developed algorithmically, provide a number of numerical examples on actual S&P 500 data, and indicate how to interpret the relevant structure from the analysis. The S&P 500 data, in particular, shows quiet, "random" periods lying between short events of strong activity; the numerical techniques presented here decompose this structure nicely.

From this analysis the authors draw one final conclusion, which is both important and somewhat surprising. That is, that the structure observed in the S&P data would seem to imply that the primary dynamics of the system is *not* driven by external "shocks", but rather is more indicative of a dynamical system which is internally driven, and exhibiting "intermittent" behavior. Such a result, if true, gives credence to the development of a dynamical approach to market analysis, and should help stimulate much work in this direction.

6 Chaos in Our Future?

The final chapter of this book provides a very unusual and speculative paper, entitled "Messy Futures and Global Brains", which is contributed by Mayer-Kress. Here, we move to the arena of socio-political-military modeling and assessment, and specifically ask how computerization, world-wide information networking, and the advent of the "new world order" could change the social and political dynamics of the future world. The author's basic premise is that two strong trends have developed since the collapse of the communist East Bloc: first, that the bi-polar, Superpower-dominated world is rapidly decentralizing into a multi-polar, factionalized system with each element pursuing strong self-interests. Secondly, that for the first time in history an information system is developing (i.e., the global computer networks, such as the Internet) which can provide the simultaneous transfer of information to all participants, i.e., a globally connected system. Together these two trends could result in a new social dynamics which has many analogies to neuronal systems (i.e., that of many simple, individual cellular units coupled in a global fashion), and hence we may see the advent of a "Global Brain" of interconnected computers and human elements. Obviously, such a system can easily display very complex dynamics, as opposed to the simpler zero-sum-games of the bi-polar political world, hence "messy" futures may become the norm, making assessment and prediction much harder.

The author consequently makes the case that the new fields of nonlinear dynamics and Complexity are the natural tools for prediction, planning, and modification of such future systems. He points out that large coupled systems such as the human brain typically operate by internally modeling reality, then choosing its more mundane actions based on the predictions of the model; on a simpler level this is termed "habituation". For large systems where enormous amounts of information are generated daily, and where time scales of reaction may be too fast for in-depth analysis, such a model-predict-feedback scheme may be necessary. He argues that dynamical systems theory has already developed many such tools for deterministic modeling and adaptive control, and that these can already be directly applied. Even more important for governments is the assessment of the *predictability* of a given scenario from such schemes, to avoid political situations which cannot be controlled, and again the tools of Complexity will come into play. Although this paper is not particularly technical, it does offer some profound insights into how chaos and Complexity theory are interpreted in such global political and social systems. In particular, the author presents an excellent intuitive discussion explaining the concepts of chaos in terms of everyday experience. He also outlines the general steps that would be necessary to construct a global information-human-computer system, that could be used as a world macro-model. What is even more fascinating is that the author makes the case that such a system could easily be constructed in the near future, hence we may be faced with this reality shortly. Finally, he closes by speculating whether other life forms (e.g., whales) may eventually be included in this world model, leading to a "global cyber-space shared by all intelligent beings on this planet". All in all, this paper provides a fitting final chapter to the monograph, and connects many of the themes of the other chapters in an intuitive and futuristic fashion.

The editors hope that the summary given in this introduction has provided a useful background and unifying theme to the contributions in this volume. As stated before, the original intention of the monograph was threefold: to introduce the latest concepts in the area of dynamical prediction and the limits of predictability; to show how these same unifying concepts are being applied in many different fields; and finally to provide the reader with sufficient technical information and numerical examples to at least begin utilizing some of the techniques in practial analyses. While the simultaneous accomplishment of all these goals seemed quite ambitious, and was by no means meant to be encompassing of the field, we feel that we have succeeded to a sufficient degree to provide a useful text. We hope that the readers will agree, particularly those who are attempting to learn these concepts for the first time. In any event, even the professional researcher should find the collection of references to be generally useful.

As a final note, we should make one caveat: that it is still not clear whether nonlinear dynamics will be generally applicable to the real world, at least not in its present simplicity. All of these methods to some degree assume that the physical systems in question have *low-dimensional* dynamical generators or representations; as yet, it is not known how realistic an assumption this may be for the Universe as a whole. Whether or not nonlinear dynamics provides a true revolution in the physical sciences, or merely gets Mankind a bit farther along the evolutionary trail, even the most brilliant dynamicist cannot predict.

References

Baake, E., Baake, M., Bock, H., and Briggs, K. (1992): "Fitting Ordinary Differential Equations to Chaotic Data". Phys. Rev. A **45**, 5524.

Casdagli, M. (1989): "Nonlinear Prediction of Chaotic Time Series". Physica D **35**, 335.

Cremers, J. and Hubler, A. (1987): "Construction of differential equations from experimental data". Z. Naturforschung **42**, 797.

Crutchfield, J. and McNamara, B. (1987): "Equations of motion from a data series". Complex Systems **1**, 417.

Farmer, J.D., and Sidorowich, J.J. (1987): "Predicting chaotic time series", Phys. Rev. Lett. **59**, 845.

Kadtke, J., and Brush, J. (1994): "Global Modeling of Chaotic Time Series with Applications to Signal Processing". *AIP Conf. Proc.* **296** (Ed. R. Katz), 205.

Kadtke, J. (1995): "Classification of Highly Noisy Time Series Using Global Dynamical Models". Phys. Lett. A **203**, 196.

Kadtke, J., and Kremliovsky, M. (1995): "Detection and Classification of Signals Using Global Dynamical Models, Part I: Theory". To appear in *Proc. of 3rd ONR Tech. Conf. on Broadband Signal Proc.*, Mystic, Conn.

Kravtsov, Yu.A. (1989): "Randomness, Determinateness, and Predictability". Sov. Phys. Uspekhi, **32**(5), 434.

Kravtsov, Yu.A., ed. (1993): *Limits of Predictability*. Springer series in Synergetics **60**, Springer-Verlag, Berlin, Heidelberg, 252.

Kravtsov, Yu.A. (1993): "Fundamental and Practical Limits of Predictability". In *Limits of Predictability*, Springer series in Synergetics **60**, Springer Verlag, Berlin, Heidelberg, 173–203.

Kravtsov, Yu.A., Etkin, V.S. (1981): "On the Role of Fluctuations in Autostochastic System Dynamics; Limitation of Predictability Time and Weak Periodical or Bits Destroying". Izv. VUZ Radiofizika **24**, 992–1006 [transl. in Radiophys. Quantum Electronics (1981) **24**, 679].

Kravtsov, Yu.A., Etkin, V.S. (1984): "Degree of Dynamical Correlation and Revealing of Dynamical Nature of Random-like Processes", Radiotechnika i Elektronika **29**, 2358–2372 [transl. in: Sov. J. Commun. Technol. Electron.(1985) **30**, 1].

Kremliovsky, M. and Kadtke, J. (1995): "Detection and Classification of Signals Using Global Dynamical Models, Part II: Experiment". To appear in *Proc. of 3rd ONR Tech. Conf. on Broadband Signal Proc.*, Mystic, Conn.

Ligthill, J. (1986): "The Recently Recognized Failure of Predictability in Newtonian Dynamics". Proc. Royal Soc., London A **407**(1832), 35–50.

Paulus, M.P., Kadtke, J.B., and Menkello, F. (1993): "Statistical Mechanics of Biological Time Series: Assessing Geometric and Dynamical Properties". Inter. J. of Bifurcation and Chaos **3**(3), 717.

Shebehely, V., Taply, B.D., eds. (1976): *Long Term Predictability*. Reidel, Dordrecht, 360.

Part 2

Time Series Analysis:
The Search for Determinism

Method to Discriminate Against Determinism in Time Series Data

Robert Cawley, Guan–Hsong Hsu[1] and Liming W. Salvino

Abstract

We describe a general, systematic method for assessing the presence or absence of determinism in time series. Our method is rooted in the standard engineering paradigm of hypothesis testing. Our application of this procedure is novel, however, for we test given data sets against the class of data sets that produce smoothness. That is, our null hypothesis is that of determinism. We highlight two inherently interactive key features of our approach which conspire to make this treatment promising, the use of a smoothness detector and of chaotic noise reduction.

1 Introduction

We remind the reader that chaos is deterministic and randomness is not deterministic. This is an inherent part of the subject.

The case of mixtures of the two, deterministic processes contaminated by random noise, and the case of randomness, or *effective randomness*, are the two necessarily presented by nature and by experiment. In the problem of distinguishing between these two distinct cases, it is therefore useful, possibly even necessary, to perform chaotic noise reduction.

It seems prudent to add the following *caveat*. As is well-known, some deterministic systems may be high dimensional, and appear pretty noisy. If they have too many degrees of freedom, it may be necessary or even best to regard them as effectively random for all practical purposes. Indeed the effort to use the data analysis methods developed for nonlinear dynamics only makes sense when the system is low dimensional. Moreover, as a practical matter, any test for determinism or any dynamical analysis and measurements performed for a given time series, have to be performed in relation to some chosen dimension, such as that used to construct the phase portraits. Conclusions regarding determinism or randomness are therefore *inherently limited by this representing dimension.*

We describe our basic philosophy and outline our two-pronged strategy for implementing it in this section and the next. A new circumstance, the availability of a smoothness detector, has made the method we describe possible in a novel and practical way, which we will explain in Sect. 2.1.

---

[1] ONR Postdoctoral Fellow.

In Sect. 3 we recall the current well-known circumstance, that the discovery of chaos has created both new opportunities and new perils, and with that a need for limits. We review the modern history of the problem of detecting dynamics in complicated time series in Sect. 4. We give over the discussion to technical matters in Sects. 5 and 6, where first we describe the smoothness detector and its workings, then go on to describe the effect of noise on a phase portrait, key elements of the workings of chaotic noise reduction, and the measured effects on smoothness.

We finish up in Sect. 7 with a summary of our recommended method: discriminating against determinism in time series data. We include a flow diagram summarizing the method and depicting its main structural features.

1.1 Our Basic Philosophy

The practical and beneficial utility, not to mention the intellectual integrity, of science depends upon a clear recognition of the role played by limits and restrictions.

"It is, of course, impossible to provide an absolutely definitive resolution of whether or not a given finite data set is chaotic" [Theiler, 1995]. It seems to us that this circumstance must form the backdrop to any serious effort to diagnose the presence of deterministic chaos in a time series.

We wish to add a comparably nontrivial emphasis on a normal science point of view, that a scientific hypothesis has to be falsifiable. Properly implemented, hypothesis testing can give expression to this requirement.

We maintain further that any conclusion one might frame in a given application, to be scientific, must be grounded in standards that are unforgiving. In practice, i.e., without being over-restrictive, they should be as unforgiving as one can manage.

Hence we must accept the conclusion that we are limited.

In particular, the task of concluding the presence of determinism is therefore necessarily one requiring judgment. This judgment must, in the end, be based upon an accumulation of statistical evidence, a point also stressed by [Theiler, 1995].

Moreover, the evidence should form a *consistent picture overall*; unless there are good reasons, *all* the features of determinism should be present and they should all be mutually consistent.

For instance, presumptive determinations of the number of active degrees of freedom, m, and embedding dimension, d_E, should not give values such that $d_E < m$ (see below). A good deal of checking and double-checking should be done in the process of the data analysis.

1.2 In Short

For any time series diagnostic method for determinism, we urge adherence to the following accepted scientific principles :

i)   it should give expression to requirements of falsifiability;
ii) it must be stern and unforgiving; and
iii) the basis of our judgment should be an internally consistent accumulation of statistical evidence.

It goes without saying that the algorithms we employ must be reliable, and carefully applied.

Our effort herein is to produce an approach that has a chance of meeting these requirements. We have devised a protocol structured to be less susceptible to the kind of erroneous judgment that leads to false alarms.

Our goal is to establish a basis for a clearly framed burden of proof in the question of the presence of dynamics in a time series, and place it on us, where it belongs.

2 Implementing Our Philosophy

Our time series analysis method involves two interacting key features. The first is a relatively new development, making possible a more economical use of hypothesis testing than heretofore. The second is an old, but infrequently used ingredient.

2.1 A New Opportunity

In our method we use the fact that smoothness in the phase space implies determinism in the time series. As an algorithm for detecting smoothness now exists [Salvino and Cawley, 1994], this circumstance creates a new hypothesis testing option for us. Rather than attempt to rule against a wide variety of random surrogates, our strategy instead can and will be to rule *directly against* the deterministic case itself!

Some deterministic processes can produce non-smooth phase portraits and we will miss these unless more refined tests are devised to permit expanded use of our approach. But this limitation seems innocuous to us since we expect that practically all physically interesting deterministic cases will involve smooth phase portraits.[2]

---

[2] We remark that dynamical systems can be, and typically are, classified exhaustively by their smoothness properties. But an exhaustive *and suitable* characterization of random processes does not seem to be available.

A particular advantage of our approach is that we should not have to devise various or appropriate kinds of random surrogate data against which to control.[3]

So our hypothesis on the data will be that phase portraits are non-smooth (H_1). The null hypothesis will be that of smooth phase portraits, hence determinism. Since this null hypothesis does not invoke a particular random class, we denote it by K_0, i.e., $H_0 \equiv K_0$. The scientific question will be: can we establish that the data we analyze are distinct from the class of time series producing smooth phase portraits (K_0)? i.e., that the data are random?

As our tactics thereby are rendered subtly different from those usually employed, our conclusions must have a subtly different character as well (Sects. 5.6 and 6.3).

2.2 Another Quote

Suppose we have ruled clearly against the null hypothesis, K_0, for a time series; it still might possess a deterministic part. That is, after chaotic noise reduction, the output time series might no longer be distinguishable from the smooth class, and we would have a deterministic conclusion.

"The use of noise-reduction methods is strongly recommended in any analysis of chaotic time-series data" [Kostelich and Schreiber, 1993]. This is an integral part of our approach also, although now for additional new reasons as well as the salutary reasons of the paper just cited.

In fact, using only modest tolerances for K_0, we find in controlled numerical studies that even small amounts of noisy contamination of a chaotic time series can make it quite easy to conclude H_1. This is as we would like to have it in an unforgiving test. But, paradoxically, by admitting noise-reduced versions of a data set for analysis, we actually can tighten those tolerances.

Summarizing, by examining noise-reduced versions of a time series, we can enlarge the effective class of datasets eligible for a determinism call; and we are able also to set more exacting and unforgiving tolerances on our criteria for phase portrait smoothness.

In other words, integration of the use of chaotic noise reduction into our approach provides a way for us to rule more strictly against K_0 (fewer false alarms) and, at the same time, give the matter our best shot (more detections)!

So we add another item to our list of important matters:

Integrate the use of chaotic noise reduction into the data analysis protocol.

In the deterministic case, of course, the added benefit is that estimates of dynamical quantities like system dimension (number of active degrees of

---

[3] As we shall see (Sect. 5.5), this hope may be only partially realized.

freedom), embedding dimension, various fractal dimensions, Lyapounov exponents, and the like, ought to be more reliable and accurate, i.e., when we use less noisy, cleaned up versions of the data to perform them.

3 Where Have We Been? — The Genesis of a Problem

The advent of chaos, which is deterministic, has led to a desire among researchers to detect determinism in time series. One catalyst for this is surely the simple geometric picture of the phase portrait of a low-dimensional dynamical system behaving chaotically, and the possibility of realizing this from an observed scalar time series with simple and experimentally robust embedding methods, such as the delay coordinate construction [apparently due to Ruelle: Packard et al., 1980; Takens, 1981]. This relatively new circumstance has created a celebrated opportunity. It may be that some processes previously supposed to be most appropriately modeled as random, in fact can be understood to be deterministic *in a simple way*. Thus, although the behavior of such data may complicated, sometimes a simple system dynamics reasonably may be concluded to have produced it.

Moreover, since the data are deterministic in these cases, we expect good things from the point of view of theoretical understanding, and possibly prediction.

But, not surprisingly, there is another side to the coin, i.e., there is a problem. The situation has sometimes threatened to become like that of the child who cried "wolf!" In an eager hope to discover chaos in complicated data, some parts of the research community have been inclined to find it under every rock.

Much of this earlier excess has by now perhaps been brought somewhat under control, the control of a little wisdom [Osborne and Provenzale, 1989], of a better awareness of statistical methods of data analysis [Theiler, 1991; Theiler et al., 1992], and of the facts as often they have proved themselves [Rapp, 1993].

4 A Brief History of Time Series and Chaos

4.1 Early Background

In the earliest heady days of experimental data analysis efforts by researchers, there was one clear and quick method of "finding chaos", viz. measure it! Measure the fractal [Mandelbrot, 1974] dimension, e.g. the box-counting dimension, d_0 [but see Greenside, 1982], or the correlation dimension, d_2 [Grassberger and Procaccia, 1983], of the phase portrait living in the reconstructed phase space of the data.

If the process is deterministic, these fractal dimensions cannot be larger than the number of active degrees of freedom — also known as the (effective) topological dimension, — m, for a minimal system of equations needed to govern a process that can produce the given data; $i.e.$ $d_F < m$, where we allow d_F to stand for a generic fractal dimension, such as d_0 or d_2.

Of course, that a naive simple measurement of a fractal dimension does not really mean one has found chaos has been generally understood, but there did appear to be a way to fix this problem. In measuring the fractal dimension, one was to vary the embedding trial dimension, d, increasing it from some lowest value, like $d = 2$.

Owing to projection effects, the fractal dimension should at first be a little low, for low values of d. But then it should rise and level off to the correct value as d is increased, and once a true embedding has been achieved.

But there is a loophole. Maybe random data could produce low fractal dimension measurements, even when this strategy is used. The well-known answer now is that they can and do [Osborne and Provenzale, 1989; Theiler, 1991].

Unhappily, though also not very much noted, this still only tells us that a hypothetical effective topological dimension, m, isn't any *smaller* than something, not that it isn't any *bigger* than something, which arguably would be somewhat more valuable.

Nonetheless, this kind of result has been widely supposed to constitute a *hint* regarding existence of an underlying dynamical simplicity.

4.2 More Looking for Hints

So, this method of looking for hints has had both problems and diseases. What is one to do, give up? Of course, not.

Other methods to "search out" chaos have appeared since the earliest days of a little more than a decade ago. Until only recently the broad idea of all of them has been the same — exploit the fact that if the data are chaotic this circumstance has textbook consequences, and those consequences can be quantified.

For instance, a given dynamical system exhibits a spectrum of finitely many, specifically m, (true) Lyapounov characteristic exponents, $\{\lambda_i \mid i = 1, \ldots, m\}$, which hopefully can be measured. Measure them.

Leaving aside the considerable technical difficulties of doing this, an important shortcoming of this method has been the occurrence of spurious exponents. In practice one necessarily finds exactly d exponents, not m. But d is a number at our disposal and therefore cannot be a property of the system producing the data.

4.3 Sharper Methods

A resolution to the last difficulty lies in the fact that it appears possible to distinguish the true exponents from the spurious ones by performing simultaneous measurements on the time reversed data [Parlitz, 1992]. This looks pretty good. The number of true exponents ought to be m, and that *is* an honest-to-god data observable.

Another good-looking example is this. With sufficiently large values of d, all evidence of transverse self-intersections should disappear in the phase portraits. Alternatively, there is a dimension $d = d_E$, the so-called "embedding dimension", above which no more "false nearest neighbors" [Kennel, Brown and Abarbanel, 1992] appear.

The occurrence of false nearest neighbors is a projection effect caused by "viewing" the geometric object constituting a true phase portrait in too low a dimension, i.e., too low from the vantage point of some (correctly) high dimension. Measure this, measure d_E.

We know that an embedding is impossible unless $d_E \geq m$. Otherwise self-intersections, cusps, or the like must occur. Thus measuring m and d_E, and barring error, we will have a result that a dimension measurement, m, is less than something, d_E. The virtue of this is that it sets an upper bound on the number of active degrees of freedom implicit in the data.

We turn now to description of the two main technical elements of our proposed approach.

5 Smoothness Implies Determinism

The material in this section is an elaboration of [Salvino and Cawley, 1994]. It was presaged by earlier work by [Kaplan and Glass, 1992] and [Wayland et al., 1993].

A smooth dynamical system is a smooth map or flow of points on a smooth manifold (phase space), which for simplicity we may suppose to be an m-dimensional Euclidean space.[4] The connection between this formal concept and an experimental observation is assumed to be a smooth mapping from the phase space to the real numbers, which are the observations. This measurement mapping may produce a quantity that is either scalar or vector valued.

In this section we describe a new method, a many vector-field method, designed to characterize and detect that smoothness when it is inherent in a data set.

---

[4] The results we describe apply to smooth dynamical systems on any smooth manifold.

5.1 A Useful Class of Vector-Field for Deterministic Time Series

Let the m-dimensional vector time series $x(t) = (x_1(t), x_2(t), \ldots, x_m(t))$ represent m simultaneous measurements from a dynamical system at time t, where $t = 1, 2, \ldots, N$. From $x(t)$ we construct a vector time series

$$V(t) = \sum_{r=0}^{R-1} c_r x(t+r).$$ (1)

$V(t)$, in fact, may be regarded as specifying a vector *field* on the phase space via $x = x(t)$; viz., $V(t) \equiv V(x(t))$. We choose $\{c_r\}$ to be a set of constants independent of x and t for simplicity. R is also a chosen constant.

In the case of flows where $x(t)$ is a solution of a system of m first order ordinary differential equations, every point $x(t)$ is distinct and nearby points in phase space behave similarly under time evolution. This is the smoothness property of a dynamical system. It implies uniqueness of solution — determinism in time series [Salvino and Cawley, 1994]. For the case of maps, also see the latter reference.

V represents the dynamical process of x on a single trajectory and it carries the smoothness. This can be illustrated by a simple example. If we choose $R = 2$, with $\{c_r\} = \{-1, 1\}$, then $V = x(t+1) - x(t)$. Here the vector field points from $x = x(t)$ to $x(t+1)$; for finely sampled flows V approximates the tangent to the trajectory at x (see below).

But suppose now only a single scalar observable $x(t)$ is available, a situation encountered in many experiments. To reconstruct the vector time series from observed data we use the delay coordinate construction of Ruelle [Packard et al., 1980; Takens, 1981], which preserves all the smoothness properties,

$$x(t) = (x(t), x(t+\Delta), \ldots, x(t+(d-1)\Delta)),$$ (2)

where d is the dimension of the embedding space and Δ is the delay time. We assume d is large enough to achieve an embedding, i.e. $d \geq d_E \geq m$, where d_E is the minimum dimension needed. The length of the reconstructed vector time series is $N = N_D - (d-1)\Delta$, where N_D is the length of the scalar time series. The vector time series in Eq. (1) becomes

$$V(t) = \sum_{r=0}^{R-1} c_r \Big(x(t+r), x(t+\Delta+r), \ldots, x(t+(d-1)\Delta+r) \Big).$$ (3)

5.2 Properties of V

V is specified by the choice of the $\{c_r\}$, but in practice its values computed from a scalar time series also depend on the embedding parameters d and Δ.

We compute V in small regions of phase space. These regions are defined by coarse graining with a uniform grid comprised of numerous small boxes of volume ε^d. For instance, if we choose the embedding dimension $d = 3$, the phase space is divided into small equal-sized boxes of size ε^3.

We introduce the directional element (unit vector) field for $\boldsymbol{V}$,

$$\hat{\boldsymbol{v}}(t) = \frac{\boldsymbol{V}(t)}{\|\boldsymbol{V}(t)\|}, \tag{4}$$

and we compute the local average of this quantity over the jth box,

$$\boldsymbol{Y}_j = n_j^{-1} \sum_{i=1}^{n_j} \hat{\boldsymbol{v}}(t_i), \tag{5}$$

where t_i is the time value for the ith point in the box. $\|\boldsymbol{Y}_j\| = 1$ if all unit vectors $\boldsymbol{v}$ are parallel in box j. If $x(t)$ is the output of a system of differential equations or maps, the unit vectors $\boldsymbol{v}(t_i)$ all will have sensibly the same orientation over the box as long as ε is small.

We compute Y_j for Lorenz system data ($\dot{x} = 10(y - x)$, $\dot{y} = 28x - y - xz$, $\dot{z} = -\frac{8}{3}z + xy$, sampling time $\Delta t = 0.05$ (17 points/cycle)), and Hénon map data ($x' = 1 - 1.4x^2 + y$, $y' = 0.3x$), for illustration (Fig. 1). The vector time series were computed from the delay coordinate construction on the x-coordinate in each case.

The choice $\{c_r\} = \{-1, 1\}$ produces a mean directional element field Y_j that has approximately unit magnitude at each box, and "arrows" varying smoothly from point to point, pointing always to the position of the next iterate. For finely sampled flows this smooth vector field approximates the flow line tangent vector field.

If the time series is generated from a random system, however, instead of varying smoothly, the direction of the Y_j formally realized under Eqs. (1–5) will fluctuate irregularly from box to box. More importantly, the corresponding individual $\hat{\boldsymbol{v}}(t_i)$ are almost surely not parallel in a given box, j, and $\|\boldsymbol{Y}_j\| \ll 1$ ought typically to result.

5.3 Determinism Tolerance

We average the mean local directional element length estimate based on (5) over the entire collection of occupied boxes,

$$W = N^{-1} \sum_j n_j \|\boldsymbol{Y}_j\|^2. \tag{6}$$

Given a choice of $\{c_r\}$, we fix d and consider the Δ dependence of W. Since, for any choice of vector field, $\|\boldsymbol{Y}_j\| = 1$ for each box j, $W = 1$ should hold for all Δ in the smooth case. However, due to finite numerics, W is often a lot less than 1. Moreover, W also depends significantly on the choice of vector field, i.e., on $\{c_r\}$.

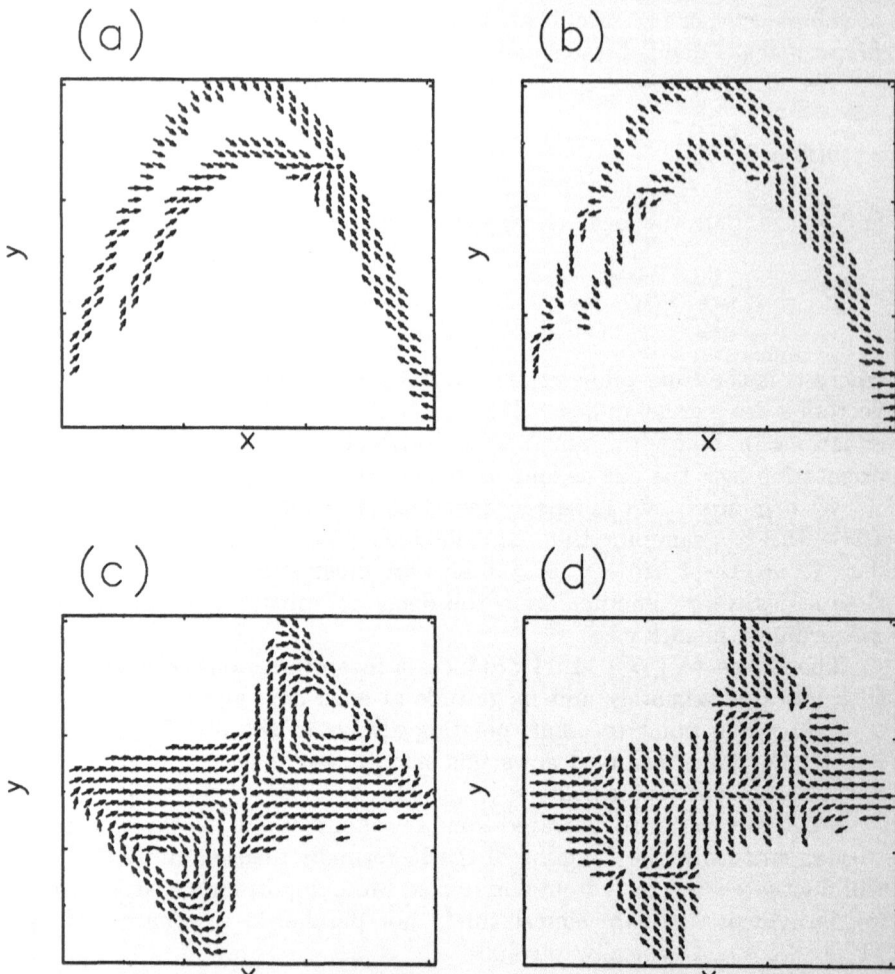

Fig. 1 Directional element fields $\boldsymbol{Y}_j$ for $d = 2$ for Hénon map (delay $\Delta = 1$) and Lorenz system ($\Delta = 4$). Data length $N_{\mathrm{D}} = 20000$, grid size $\varepsilon^2 = 1/30 \times 1/30$. (a) Hénon: $\{c_r\} = \{-1, 1\}$. (b) Hénon: $\{c_r\} = \{2, -5, 3\}$. (c) Lorenz: $\{c_r\} = \{-1, 1\}$. (d) Lorenz: $\{c_r\} = \{2, -5, 3\}$.

But this last circumstance can be exploited, for there is a very wide range of options available to us from the arbitrariness of the $\{c_r\}$ in (3).

Basically, we look for the largest computed values of W we can get. We know that $W = 1$ can never be achieved, only something like $W \simeq 1$. So for practical applications we need to set a "determinism tolerance limit", i.e., a limit on the error we will accept in allowing $W \simeq 1$ rather than demanding $W = 1$, exactly.[5]

In general, we feel that it is useful to make the test stringent and unforgiving so that we will less likely be fooled by false alarms. Somewhat arbitrarily, we now take our limit to be 0.9. That is, if $W > 0.9$, we will accept this as meaning that $W \simeq 1$. This will mean then that the phase portraits are smooth, and that there is therefore clear evidence of determinism present.

5.4 W_M, a Stable Index of Smoothness

We list ten vector fields in Table 1; for convenience we have chosen $\sum_{r=0}^{R-1} c_r = 0$. We compute $W(\Delta)$ for each vector field; and for each delay Δ we further identify both maximum and minimum values of W, viz. $W_M(\Delta)$ and $W_m(\Delta)$, over the ten choices. The results are shown in Fig. 2.

Our computational parameters are $d = 3$, $N_D = 20000$, and our linear grid size, set by the normalized maximum range of the data, is $\varepsilon = 1/40$. Except where stated, we use these values in this section. The range of Δ for the Lorenz data, up to $\Delta = 3000$, corresponds to about 180 cycles.

Table 1. Coefficients c_r for ten vector fields V_n, $n = 1, 2, \ldots, 10$, used in computations of $W(\Delta)$

	V_1	V_2	V_3	V_4	V_5	V_6	V_7	V_8	V_9	V_{10}
c_0	−1.0	−3.0	2.0	4.7	−2.0	3.5	−3.4	1.0	0.9	3.0
c_1	1.0	4.0	−5.0	−3.0	3.0	−2.7	−0.5	2.0	0.8	−2.0
c_2	0.0	−1.0	3.0	−1.7	−4.0	−1.4	−0.1	3.0	−3.5	0.0
c_3	0.0	0.0	0.0	0.0	3.0	0.6	4.0	−4.0	4.0	2.0
c_4	0.0	0.0	0.0	0.0	0.0	0.0	0.0	−2.0	−2.2	−3.0

Although the differences between values of $W_M(\Delta)$ and $W_m(\Delta)$ are systematic and large, as Δ rises the *upper envelopes* of the $W(\Delta)$ plots descend to well-defined constant values (Fig. 2). We denote these values by W_M for the maximum and W_m for the minimum.

In the examples given in Fig. 2, we have found $W_M \simeq 1$ for both Lorenz ($W_M = 0.92$) and Hénon ($W_M = 0.99$) time series. Since there are a number of factors that may contribute to the finite numerics problem, such as the choices of vector field, grid size, embedding dimension d, input time series length, and sampling rate, we feel that $W_M > 0.9$ is a strict requirement.

---

[5] Strictly, we should call it a smoothness tolerance since that is what W_M measures.

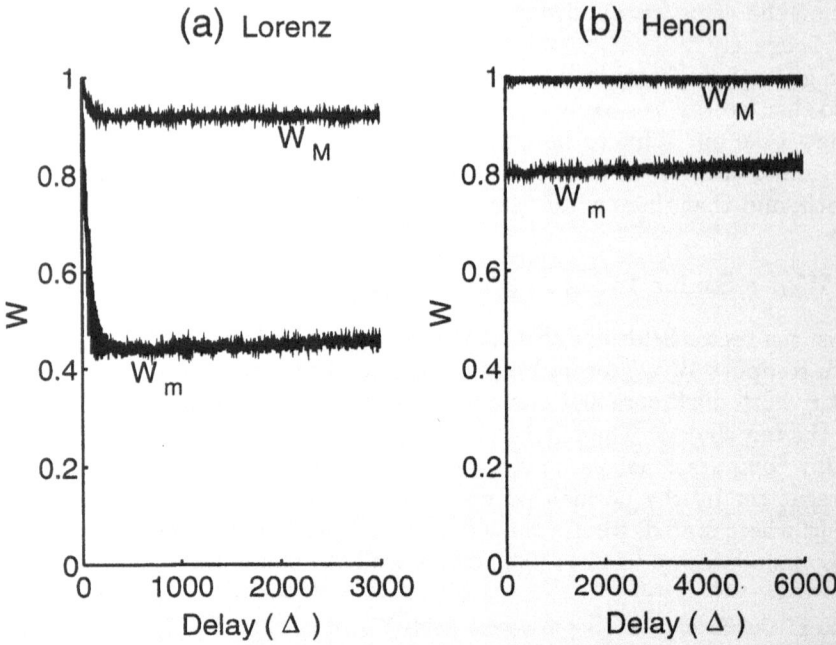

Fig. 2 Determinism case studies for (a) Lorenz and (b) Hénon time series. Computed W values shown are maxima (W_M) and minima (W_m) for each delay over the ten arbitrarily chosen vector fields in Table 1.

In fact, standard chaotic datasets sometimes fail to measure up to this requirement. The Ikeda map[6] is one example of this, where we found $W_M = 0.87$, which is high, but not high enough for our requirement. We found $W_M = 0.89$ when we raised the embedding dimension for the computation to $d = 4$, so $W_M \simeq 1$ fails in both cases.

Another example where $W_M \simeq 1$ fails is a relatively low noise laboratory experimental time series that is known to be chaotic. The data represent the horizontal displacement of the base of a magnetostrictive ribbon in the experiment of [Ditto et al., 1989]. Our measurement gave $W_M = 0.85$, which again is a high value.

Evidently, these results do not necessarily mean that $W_M \simeq 1$ cannot be achieved for the dataset. For example, some wider choice of vector fields might succeed and a simple thing to do would be to enlarge the set used for the computation.[7]

---

[6] Our Ikeda map [Hammel, Jones and Moloney, 1995] time series was $x(n)$, where
$$z(n) = x(n) + iy(n), \quad z(n+1) = 1.0 + 0.9z(n) \exp\left[0.4i - \frac{6.0i}{(1+|z(n)^2|)}\right].$$
[7] We tried this for these examples, but it didn't work.

It would seem that, unless we ease our choice of determinism tolerance, many chaotic datasets may, unfortunately, be numerically distinguishable from the smooth class. We address this problem in the next section.

5.5 Comparison Test — Making Use of W_m

We would like to accommodate high values of W_M that may indeed be a signal of determinism, but fail to meet the strict requirement, $W_\mathrm{M} \simeq 1$. At the same time we also want to avoid relaxing the tolerance standard.

Accordingly, we have devised a fall-back position, one allowing us to consider W_M values that are clearly high, but not as high as 0.9. If $0.7 < W_\mathrm{M} < 0.9$, we write $W_\mathrm{M} \sim 1$, and consider how we might proceed in such cases.

We use a method of surrogate data sets [Theiler et al. 1992]. Our method is a simple comparison test that can distinguish the given data against a selected class of random surrogates, which will become the null hypothesis, H_0. The surrogate class may be specified in any of a number of ways [Theiler et al., 1992]. We use again the arbitrariness in the choice of vector field V.

For the examples below we generated surrogate data from the Fourier transform (periodogram) of the given data by randomizing the phases and transforming back [Algorithm I in Theiler et al., 1992]. Using the vector fields in Table 1, we then computed both W_M and W_m for the surrogate data and for the given data.

Organizing the results as shown in Fig. 3 for the Ikeda map, we easily distinguish the given data from the surrogate. For suppose in Fig. 3 the Ikeda map time series had been some other realization of the surrogate. Panels (a) and (b) would look exactly alike. This was actually the case for ambient ocean acoustic sonobuoy data (Fig. 4). Thus, using this method, the acoustic data plots show little evidence of smoothness and therefore determinism, while the Ikeda map data show strong evidence of determinism.

When we applied this test to the ribbon dataset it also was clearly distinguishable from the surrogate, so here again the evidence for determinism is strong [Salvino and Cawley, 1992].

5.6 Not Random at All?

From this point of view we regard $W_\mathrm{M} \sim 1$ as providing evidence for determinism if it can be supported by a positive result from the comparison test — that is, if the given data set can be distinguished from one belonging to a random surrogate class.

We can strengthen (or refute!) our basis for a determinism call in the $W_\mathrm{M} \sim 1$ case by implementing it repeatedly against a variety of surrogate classes. We agree with the emphasis of [Theiler et al., 1992], that this should be an important part of a serious protocol of "chaos hunting". In this way we can build the case for concluding that a data set cannot be distinguished from the smooth class (K_0).

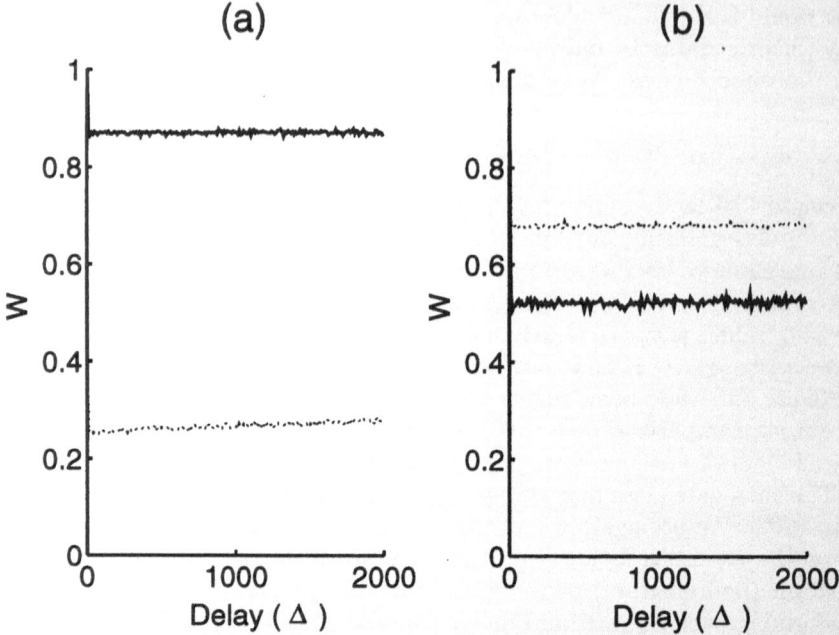

Fig. 3 W_M and W_m plots for comparison test analysis of Ikeda map data: solid — Ikeda data, dotted — surrogate data. For each Δ, $W =:$ (a) maximum for data, minimum for surrogate; (b) minimum for data, maximum for surrogate.

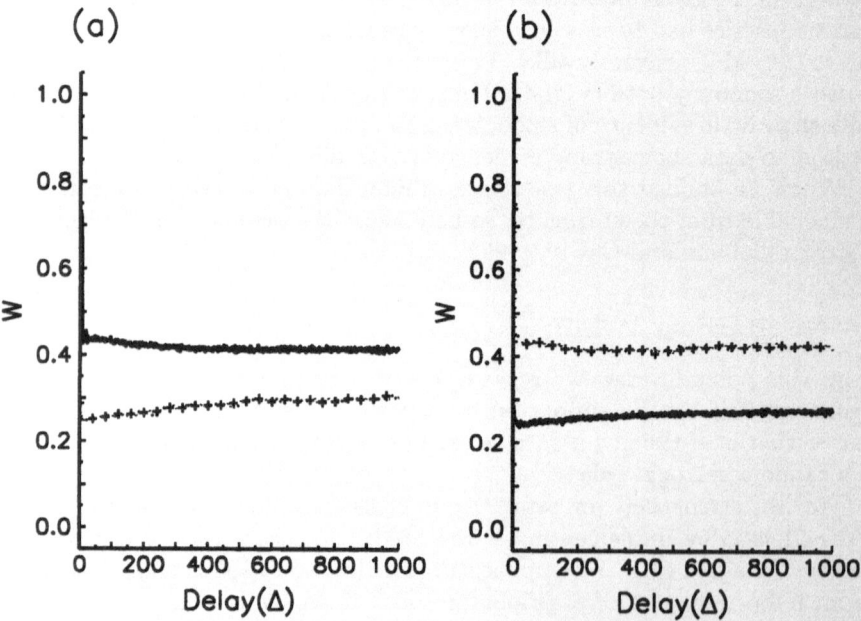

Fig. 4 W_M and W_m plots for comparison test analysis of ambient ocean acoustic data: solid — given data, dotted — surrogate data. For each Δ, $W =:$ (a) maximum for data, minimum for surrogate; (b) minimum for data, maximum for surrogate.

Assuming we have done this, the resulting conclusion is still a little weaker than that issuing from $W_M \simeq 1$; for the latter means the data set cannot be distinguished from the smooth class (K_0) at all. That is, it is consistent with deterministic in the sense of being "not random at all". There ought not be some surrogate out there that we didn't think of in our comparison testing.

We are still not done, however, for our dataset may have enough noise to mask the presence of determinism.

6 Effects of Noise

Successful application of a chaotic noise reduction algorithm can recover a phase space smoothness lost under the influence of noise. We describe the basic background of the effects noise and of the operation of chaotic noise reduction algorithms, and we give a few results about what happens to the W_M measurements.

6.1 The Phase Portraits

In this section we consider time series that are both noisy and chaotic. We assume the noisy part is some form of randomness. In principle the noise may be either instrumental (additive) or dynamical.

We now denote the *given* dataset by $x(t)$, $t = 1, 2, \ldots, N_D$; without loss of generality, we may write

$$x(t) = X(t) + \varepsilon\eta(t), \ t = 1, \ldots, N_D, \tag{7}$$

where $X(t)$ is now the underlying deterministic part. We have written the random part of the time series as $\varepsilon\eta(t)$, where $\eta(t)$ is a zero mean process having unit variance. Like typical forms of noise, chaotically generated time series have infinite bandwidth. Consequently, conventional band-passing methods for noise reduction can lead to undesirable effects, which noise reduction methods based on dynamical systems theory can avoid. See [Kostelich and Schreiber, 1993] and [Grassberger et al., 1993] for recent surveys.

When the time series $x(t)$ represents a (noise-free) dynamical system the tip of the delay vector x in (1) traces out a geometrical object, the embedded image of the attractor in $\mathbb{R}^d$. For a chaotic time series this object is a fractal, but the flow lines, unstable manifolds, and dynamics all are smooth.

Note also that the noise-free attractor image is contained in, and samples the m-dimensional embedded image of the original true space for the dynamics, $\mathbb{R}^m$. But when the data are noisy as in (7) the orbit points in $\mathbb{R}^d$ fluctuate out of this $\mathbb{R}^m$,[8] resulting in a "fuzzy" and not-so-smooth image of the geometrical object.

---

[8] Precisely, they fluctuate out of the embedded image, $(\mathbb{R}^m)' \subseteq \mathbb{R}^d$, of $\mathbb{R}^m$.

One result of a successful application of chaotic noise reduction to a data set will be to recover an underlying smoothness that would be there were it not for the noise.

6.2 Chaotic Noise Reduction

For the last claim to be correct, of course, the underlying smoothness actually must be present.

Evidently, when a raw data set is given we do not have the decomposition in Eq. (7) available. For a noise reduction method to be meaningful (and successful!), it must correctly exploit some aspect of dynamical systems theory to provide a basis of identification of a deterministic part.

A variety of different noise reduction methods exist [Kostelich and Schreiber, 1993; Grassberger et al., 1993], differing from one another according to which aspect is exploited. Typically, these proceed through an identifiable sequence of four steps.

Step 1. Embed; i.e., replace the given scalar dataset, $x(t)$, by a data-state vector, such as $\boldsymbol{x}(t)$ in Eq. (7). This provides a phase portrait in $\mathbb{R}^d$ for the manipulations of noise reduction.

Step 2. Adjust the positions of the points of the phase portrait. This is what is normally regarded as the key noise reduction step, with $\boldsymbol{x}(t)$ replaced by $\hat{\boldsymbol{x}}(t)$.

Step 3. Disembed; i.e., replace the altered data–state vector time series, $\hat{\boldsymbol{x}}(t)$, by a new scalar dataset, $\hat{x}(t)$. This step can remove further noise [Cawley and Hsu, 1992].

Step 4. Iterate the foregoing, inputting $\hat{x}(t)$ in Step 1.

Sometimes the order of these steps is shuffled. For instance, in the trajectory adjustment step (Step 2) [Kostelich and Yorke, 1988] iterate before disembedding. In their method one can do it either way in principle, although that is not always the case.

We remark, in passing, that iteration is an essential part of chaotic noise reduction since the process is one of successive approximations. Time series improvements otherwise will be small. This circumstance has significant and useful consequences involving attractor stability and instability under iteration [Cawley and Hsu, 1992b], and quantitative measures of such effects [Cawley, Hsu and Salvino, 1994]. This creates opportunities for our program, which space does not permit our going into here.

6.3 Smoothness After Noise Reduction

Even low levels of noise can interfere with the determinism test.

We applied the LGP algorithm to the ribbon data, where the noise is dynamical with $SNR_i \sim 35\,\text{dB}$. We computed W_M using the vector fields in

Table 1. This time we got $W_M = 0.91$, so that $W_M \simeq 1$ for the noise-reduced version. For comparison, we recall the value, $W_M = 0.85$, found in Sect. 5.4 for the raw data set.

So the (noise-reduced) ribbon data now are consistent with "not random at all".

As a control we subjected the Ikeda map dataset to noise reduction. We found that W_M was unchanged, giving again only $W_M \sim 1$.

Noise reduction had no discernible effect on the ambient acoustic noise time series, and W_M again obeyed a clear $W_M < 1$.

Finally, another example of an experimental dataset we examined is a time series ($N_D = 12\ 000$) consisting of intensity measurements of a laser in a chaotic state. The data were reported and analyzed by [Hübner et al., 1993]. We found $W_M = 0.83$, so we can only say $W_M \sim 1$.

We applied the noise reduction algorithm to the laser data. W_M only increased slightly and still lay in the range $W_M \sim 1$. So here we were unable to conclude consistency with "not random at all".

7 Summary and Discussion

We summarize our philosophy of approach to a data set in a flow chart (Fig. 5). When we get to the bottom of the chart, whether the end result has been conclusive or inconclusive, it is vital still to go back and retest noise-reduced versions of the data.

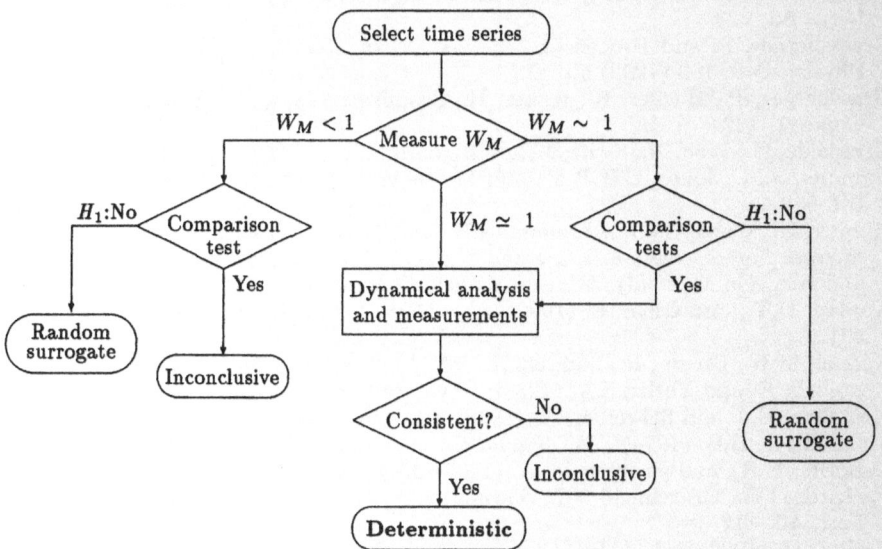

Fig. 5 Flow chart for time series analysis to discriminate agains smoothness.

Suppose our analysis has brought us to a conclusion that our data, or a noise-reduced version of our data, are indistinguishable from the smooth class so that we do want to go further with the analysis. There are many things we *should* then do, there are many excellent ideas in the literature now. A recent reference containing several of these in conjunction is [Tuffilaro et al., 1995].

As we have said, no method can be better than its algorithms — in the present case, the smoothness detector and noise reduction algorithm. What we have done in this article is to advocate a stern procedure, but it is only a procedure for getting started, for finding a basis for a go/no-go decision in the analysis of a data set.

We think this should always be the first step taken.

References

Cawley, R. and Hsu, G.-H. (1992a): Phys. Rev. A **46**, 3057.

Cawley, R., and Hsu, G.-H. (1992b): *Another new chaotic noise reduction algorithm*, in Proc. of the First experimental chaos conference, Arlington, VA, October 1–3, 1991 (ed. by S. Vohra, M. Spano, M. Shlesinger, L. Pecora and W. Ditto). World Scientific Press, Singapore, 38–46.

Cawley, R., Hsu, G.-H. and Salvino, L.W. (1994): *Detection and diagnosis of dynamics in time series data: theory of noise reduction*, in The chaos paradigm: developments and applications in engineering science, AIP Conference Proceedings Vol. 296, American Institute of Physics, New York, pp. 182-192.

Cawley, R., Hsu, G.-H., and Salvino, L.W. (1995) (in preparation).

Ditto, W.L., Rauseo, S.N., Cawley, R., Grebogi, C., Hsu, G.-H., Kostelich, E., Ott, E., Savage, H.T., Segnan, R., Spano M.L., and Yorke, J.A. (1989): Phys. Rev. Lett., **63**, 923.

Grasssberger, P. and Procaccia, I. (1983): Phys. Rev. Lett. **50**, 346 (1983); and Physica D **9**, 189 (1983).

Grassberger, P., Hegger, R., Kantz, H., Schaffrath, C., and Schreiber, T. (1993): Chaos **2**, 127.

Greenside, H., Wolf, A., Swift, J., and Pignataro, T. (1982): Phys. Rev. A **25**, 3453.

Hammel, S.M., Jones, C.K.R.T., and Moloney, J.V. (1985): J. Opt. Soc. America, B**2**, 552.

Hübner, U., Weiss, C.-O., Abraham, N.B., and Tang, D. (1993): in *Time Series Prediction: Forecasting the Future and Understanding the Past* (ed. by A.S. Weigend and N.A. Gershenfeld). Addison-Wesley, Reading, Mass. 73.

Kaplan, D.T., and Glass, L. (1992): Phys. Rev. Lett. **68**, 427); and Physica D **64**, 431.

Kennel, M.B., Brown, R., and Abarbanel, H.D.I. (1992): Phys. Rev. A **45**, 3403.

Kostelich, E. and Yorke, J.A. (1988): Phys. Rev. A **38**, 1649.

Kostelich, E.J. and Schreiber, T. (1993): Phys. Rev. E **48**, 1752.

Mandelbrot, B.B. (1974): *The Fractal Geometry of Nature*, W.H. Freeman.

Osborne, A.R., and Provenzale, A. (1989): Physica D **35**, 357.

Packard, N.H., Crutchfield, J.P., Farmer, J.D., and Shaw, R.S. (1980): Phys. Rev. Lett. **45**, 712.

Parlitz, U. (1992): Int. J. Bif. Chaos **2**, 155.

Rapp, P.E. (1993): Biologist **40**, 89.

Salvino, L.W. and Cawley, R. (1994): Phys. Rev. Letters **73**, 1091.

Takens, F. (1981): in *Lecture Notes in Mathematics* (ed. by D.A. Rand and L.-S. Young). Vol. 898, p. 366. Springer-Verlag, Berlin.

Theiler, J. (1991): Phys. Lett. A **155**, 480.

Theiler, J. (1995): Phys. Lett. A **196**, 335, 1995.

Theiler, J., Eubank, S., Longtin, A., Galdrikian, B., and Farmer, J.D. (1992): Physica D **58**, 77.

Tufillaro, N.B., Wyckoff, P., Brown, R., Schreiber, T., and Molteno, T. (1995): Phys. Rev. E **50**, 164.

Wayland, R., Bromley, D., Pickett, D., and Passamante, A. (1993): Phys. Rev. Lett. **70**, 580.

Observing and Predicting Chaotic Signals: Is 2% Noise Too Much?

Thomas Schreiber and Holger Kantz

Abstract

We discuss the influence of noise on the analysis of complex time series data. How harmful it is depends on the nature of the noise, the complexity of the signal and on the application in mind. We will give generally valid upper bounds on the feasible noise level for dimension, entropy and Lyapunov estimates and lower bounds for the optimal achievable prediction error. We illustrate in a number of examples why it is hard to reach these bounds in practice. We briefly sketch methods to detect, analyze and reduce measurement noise.

1 Introduction

All experimental data are to some extent contaminated by noise. That this is an undesirable feature is commonplace (by definition, *noise* is the unwanted part of the data.) But how bad is noise really? The answer is as usual: it depends. The nature of the system emitting the signal and the nature of the noise, determine if the noise can be separated from the clean signal at least to some extent. This done, the amount of noise introduces limits on how well a given analyzing task (prediction, etc.) can be carried out.

In order to focus the discussion in this chapter on the influence of noise, we will assume throughout that the data are otherwise largely well-behaved. By this we mean that the signal would be to some extent predictable by exploiting an underlying deterministic rule — were it not for the noise. This is the case for data sets which can be embedded in a low dimensional phase space, which are stationary and which are not too short. Violation of one of these requirements leads to further complications which will not be addressed here.

The signals we will consider in the following are typically observations of dynamical phenomena. By this we mean that they are based on a (generally unknown) low dimensional deterministic system. Thus the evolution can be expressed either as a set of ordinary differential equations or as a discrete time mapping. Although most physical systems evolve continuously in time, data are always sampled discretely. Since, furthermore, the results we will present assume a simpler form, without loss of generality, we will concentrate on maps in what follows. Thus, a trajectory of the underlying system can be obtained by iterating

$$\boldsymbol{y}_{n+1} = \boldsymbol{F}(\boldsymbol{y}_n) , \tag{1}$$

starting from some initial condition $\boldsymbol{y}_0$. Reality deviates from this description in several respects. First, we usually cannot obtain $\boldsymbol{y}_n$ directly, instead we measure some function $g(\boldsymbol{y})$ of it, in most cases yielding only scalar values. Second, this measurement is always subject to some measurement error, be it due to random fluctuations or due to the discretization. Finally, at each moment the state $\boldsymbol{y}_n$ of the system may be perturbed by a stochastic process.

For all kinds of data processing described below we have to reconstruct a higher dimensional trajectory $\{\boldsymbol{x}_n\}$ which is in some sense equivalent to $\{\boldsymbol{y}_n\}$, knowing only noisy scalar measurements $\{x_n\}$. Without noise, this is successfully achieved using a time-delay embedding [Sauer, Yorke, and Casdagli, 1991]. In the presence of noise state space reconstruction becomes a very intricate problem. However, we will cut the discussion short by assuming that a delay embedding of sufficiently high dimension forms a proper reconstruction and our data yield trajectories deterministic up to the noise. For the reconstruction problem we refer the reader to the literature, in particular to a paper by [Casdagli et al., 1991]. A review of nonlinear time series analysis can be found in [Grassberger, Schreiber, and Schaffrath, 1991].

2 Measurement Error and Dynamical Noise

Measurement noise refers to the corruption of observations by errors which are independent of the dynamics. The dynamics satisfies $\boldsymbol{y}_i = \boldsymbol{f}(\boldsymbol{y}_{i-1})$, but we measure scalars $x_i = g(\boldsymbol{y}_i) + \eta_i$, where g is a smooth function that maps points on the attractor to real numbers, and the η_i are independent and identically distributed (IID) random variables. (Even in multi-channel measurements, generally not $\boldsymbol{y}_n + \eta_n$ is recorded, but different scalar variables corresponding to different measurement functions g_j.)

Dynamical noise, in contrast, is a feedback process wherein the system is perturbed by a small random amount at each time step:

$$x_i = f(x_{i-1} + \eta_{i-1}) . \tag{2}$$

Dynamical and measurement noise are two notions of the error that may not be distinguishable a posteriori based on the data only. Both descriptions can be consistent to some extent with the same signal. [9]

---

[9] For strongly chaotic systems which are everywhere expanding (more precisely, for *Axiom A* systems), measurement and dynamical noise can be mapped onto each other [Eckmann and Ruelle, 1986].

In this article we do not want to address in detail the question of how to distinguish and possibly reduce these different kinds of noise. The noise reduction problem is discussed in some degree in the contribution of R. Cawley and G.-H. Hsu to this volume.

However, the effects of the noise on the predictability and other properties of the data will indeed sometimes depend on the nature of the noise and we will thus specify in these cases, which kind of noise we are assuming. Generally, dynamical noise induces much greater problems in data processing than additive noise, since in the latter case a nearby clean trajectory of the underlying deterministic system exists. Furthermore, what one interprets to be dynamical noise sometimes may be a higher dimensional deterministic part of the dynamics with small amplitude. Even if this is not the case, dynamical noise may be essential for the observed dynamics. Consider a situation, where the system possesses a weakly attracting periodic orbit and a chaotic repeller, for example the logistic equation $x_{n+1} = a - x_n^2$ with $a = 1.9408$ [Kantz and Grassberger, 1985]. A typical noise-free trajectory will eventually settle down to a period-three orbit. Now, if we let Gaussian noise interact with the dynamics, there is a finite probability that the noise kicks the trajectory off the periodic orbit onto the repeller. The signal in such a case will consist of chaotic bursts with exponentially distributed durations, embedded in intervals of periodic motion. If one tries to model this behavior without noise, one will presumably need a very complicated model. We do not want to enter such intricate problems, but this example should serve as a warning that dynamical noise may have more effects than simply smearing out some small-scale deterministic structures.

3 Noise and Prediction

Let us first consider pure measurement noise, i.e., the measured data x_i can be thought of as composed of a deterministic part y_i plus random errors η_i. Noise limits the accuracy of predictions in three ways: (i) The prediction error cannot be smaller than the noise level since we cannot predict the noise part of the future measurement; (ii) The values we base our prediction on are themselves noisy; (iii) In the generic case that we have to estimate the dynamical evolution from the data, this estimate will be affected by noise.

3.1 Examples with Known Dynamics

The most obvious limit of predictability is given by the fact that we have to determine the prediction error with respect to the noisy future of the signal, thus the forecasting error is never smaller than the noise level, $s^2 > \langle \eta^2 \rangle$. This would be true even if we knew everything and could predict the signal component y_{n+1} exactly; we can never predict the noise component which has variance $\langle \eta^2 \rangle$.

Now say we know the time evolution f and want to use it to predict one step ahead in time, going out from the noisy present state. Then for a one dimensional example the prediction error is defined as $s^2 = \langle [x_{i+1} - f(x_i)]^2 \rangle$, i.e.,

$$s^2 = \sum_{i=1}^{n} [y_{i+1} + \eta_{i+1} - f(y_i + \eta_i)]^2 , \qquad (3)$$

which means for small noise level

$$s^2 = \sum_{i=1}^{n} \left[y_{i+1} + \eta_{i+1} - f(y_i) - \eta_i f'(y_i) - \frac{1}{2} \eta_i^2 f''(y_i) \right]^2$$
$$= \sum_{i=1}^{n} \left[\eta_{i+1} - \eta_i f'(y_i) - \frac{1}{2} \eta_i^2 f''(y_i) \right]^2 , \qquad (4)$$

and if the noise is uncorrelated, has zero skewness, and is independent of the signal

$$s^2 = \langle \eta^2 \rangle [1 + \langle f'(y)^2 \rangle] + \frac{1}{4} \langle f''^2 \rangle \langle \eta^4 \rangle . \qquad (5)$$

In higher dimensional examples the Jacobian matrix takes the role of f' and the prediction error can be estimated as

$$s^2 \approx \langle \eta^2 \rangle [1 + \langle \Lambda(y)^2 \rangle] , \qquad (6)$$

where $\Lambda(y)$ is the largest eigenvalue of the Jacobian matrix J of F at the point y, $J_{ij} = \partial F_i / \partial y_j$. For chaotic systems $\langle f'^2 \rangle$ and $\langle \Lambda^2 \rangle$ are larger than one.

Before we discuss predictions over longer times let us outline how we could slightly improve the forecast if we not only knew the dynamics, but also the distribution of the noise, or at least the noise level.

Suppose we want to predict x_{n+1}, given x_n. The most probable outcome ignoring the noise would be $F(x_n)$. However, under the influence of Gaussian noise of amplitude σ the expectation value of x_{n+1} is

$$\hat{x}_{n+1} = \frac{1}{(\sigma\sqrt{2\pi})^m} \int d^m\eta \, F(x_n - \eta) e^{-\frac{\eta^2}{2\sigma^2}} , \qquad (7)$$

where m is the dimension of phase space. For nonlinear F, $\hat{x}_{n+1}$ is different from $F(x_n)$. In particular, for the one-dimensional quadratic map we obtain

$$\hat{x}_{n+1} = \frac{1}{\sigma\sqrt{2\pi}} \int d^m\eta \, [a - (x_n - \eta)^2] e^{-\frac{\eta^2}{2\sigma^2}} = a - \sigma^2 - x_n^2 . \qquad (8)$$

Thus the best estimate of x_{n+1} knowing the noisy x_n is not found using the exact dynamics but the effective dynamics $\hat{f} = \hat{a} - x_n^2 = a - \sigma^2 - x_n^2$.

Inserting $\hat{f}$ in Eq. (3) reduces the average forecast error by $2\sigma^4$. For uniformly distributed noise the improvement is more significant.

In order to make predictions more than one step into the future, we have to average over all possible perturbations η_i at the intermediate times $n + i$ as well in order to estimate the most probable outcome. Under realistic conditions we usually know neither the noise nor the dynamics with the accuracy necessary for this strategy.

Table 1 Limits of predictability for some typical chaotic systems. Here we assume that the dynamical equations are known but the data is noisy. The columns contain the following values, obtained by numerical simulation: (i) s/σ, the rms prediction error for one-step predictions in units of the noise level; (ii) The maximal noise level for which better than chance predictions are possible one step ahead; (iii) The maximal Lyapunov exponent [Kantz, 1994]; (iv) The maximal number of steps which can be in principle predicted with an initial noise level of 2%, following Eq. (10).

System	(i)	(ii)	(iii)	(iv)
quadratic map, $a = 2$	2.81	36%	$\ln 2 \approx 0.69$	5.6
Hénon map	1.61	60%	0.416	9.4
Ikeda map	1.80	56%	0.505	7.7

Up to now we have only studied the one-step prediction error, which is dominated by the average local expansion rate $\langle \Lambda(y)^2 \rangle$. For longer prediction times we have to use a different approach, since the error is not dominated by the permanent perturbation of any future measurement but by the deterministic evolution of the error in our "initial condition". This uncertainty in the last known value, from which we have to start our predictions, will after a few time steps align along the unstable manifold of the attractor of the system. The quantity which describes the instability along this direction is the maximal Lyapunov exponent λ. Initial errors η_{n_0} grow on average as

$$\langle \eta_n / \eta_{n_0} \rangle \approx e^{\lambda(n - n_0)} \tag{9}$$

for long prediction times $n - n_0$. For one dimensional maps the exponent λ is given by $\langle \log f'(y) \rangle$, where the average has to be taken according to the natural distribution of iterates of the map. Note that we have to perform a geometric mean instead of the arithmetic one encountered above. But not only is the growth rate different from the one above, even the qualitative behavior is different: for short times, neighboring trajectories separate diffusively, while for long times they separate exponentially.

A fundamental limit of predictability is thus given by the time needed until the initial uncertainty has grown to the size of the variance of the signal. Since the initial uncertainty is given by the noise level $\sqrt{\langle \eta^2 \rangle}$, the maximal average prediction time is

$$\Delta n_{\max} = \frac{\ln\langle y^2\rangle - \ln\langle \eta^2\rangle}{2\lambda} . \tag{10}$$

In some sense this is a simplification: Since in Eq. (9) only finite times $\Delta n = n - n_0$ enter, for the increase of a given error η_n not the (infinite-time averaged) Lyapunov exponent but the effective (finite-time) local expansion rate is of relevance. They may assume quite different values on different parts of the attractor. Thus on parts of the attractor where this rate is smaller than average one can predict farther into the future. On the other hand, there may be situations where the local expansion rate is very large such that it is nearly impossible to predict beyond such an event. This behavior is quite common for many flows, one example being the Lorenz system [Lorenz, 1963]: Only when the trajectory is close to flipping from one wing of the attractor to the other is its precise position of importance for its future, especially for the amplitude of the oscillations on the other wing. From this point of view the prediction algorithm [Meyer and Packard, 1992] is very interesting: it ignores regions of the state space where predictions are impossible.

3.2 Dynamics from a Time Series

The realistic case where the dynamical evolution $\boldsymbol{F}$ has to be estimated from the data is hard to study in general.[10] Only some simple cases can be understood theoretically, more realistic examples can be evaluated numerically. Let us investigate the ideal case first and say we know what $\boldsymbol{F}$ looks like, but need to fit some coefficient(s). As an example suppose we try to estimate the parameter a for data generated with the quadratic map $y_{n+1} = a - y_n^2$ in the presence of noise, using a least squares procedure. We minimize

$$\sum_n [x_{n+1} - \hat{f}(x_n)]^2 = \sum_n [x_{n+1} - \hat{a} + x_n^2]^2 \tag{11}$$

as usual over the available data. Due to the fact that $x_{n+1} = y_{n+1} + \eta_{n+1} = a - (y_n + \eta_n)^2$ we will obtain a biased estimate for $\hat{a}$:

$$\hat{a} = a - 2\langle y\eta_n\rangle - \langle \eta_n^2\rangle = a - \sigma^2 , \tag{12}$$

assuming the noise to be independent of the signal. The bias is introduced since the variable x_n is taken as the independent one, neglecting the fact that it is contaminated by noise. Glancing back at Eq. (8) we observe that in the case of the quadratic map the bias exactly introduces the correction needed to predict the most probable outcome. This will not be true for maps with $f''' \neq 0$ since there the optimal correction will depend on x while the induced bias only represents the average of these corrections.

A more important point is that we will get unsatisfactory results if we iterate the biased map in order to predict further into the future, since we

---

[10]However, some interesting material can be found in [Kostelich, 1992].

generate a series of most probable values, which in this sense are noise free. But applied to noise-free data the effective dynamics $\hat{f}$ leads to wrong predictions, such that after the first prediction step one should use the correct dynamics f.

Another example which can still be studied analytically and which yields quite unexpected results is the Hénon map. Given a long but noisy Hénon trajectory, let us assume that we know the functional form of the equations, $y_{n+1} = 1 - ay_n^2 + by_{n-1}$, but we do not know the actual values of a and b which have been used to create the trajectory. An ordinary least squares fit, minimizing

$$\sigma^2 = \sum_{n=3}^{N} [x_{n+1} - (1 - ax_n^2 + bx_{n-1})]^2 \tag{13}$$

produces systematically wrong estimates $\hat{a}$ and $\hat{b}$. One can after some algebra derive the following analytical result of the minimization of Eq. (13):

$$\langle y_n^2 y_{n-1} \rangle = S_y$$

$$\hat{a} = \frac{a(\sigma^2(\langle y_n^4 \rangle + \langle y_n^2 \rangle \langle \eta_n^2 \rangle)) - S_y(S_y + \langle y_n^2 \rangle \langle y_n \rangle)) - b\langle \eta_n^2 \rangle(\langle y_n \rangle \langle \eta_n^2 \rangle)}{\sigma^2(\langle y_n^4 \rangle + 6\langle y_n^2 \rangle \langle \eta_n^2 \rangle + \langle \eta_n^4 \rangle) - (S_y + \langle y_n \rangle \langle \eta_n^2 \rangle)^2}$$

$$\hat{b} = \frac{(\hat{a} - a_0)S_y + \hat{a}\langle y_n \rangle \langle \eta_n^2 \rangle + b_0 \langle y_n^2 \rangle}{\sigma^2}, \tag{14}$$

where $\sigma^2 = \langle y_n^2 \rangle + \langle \eta_n^2 \rangle$ is the variance of the noisy signal. The moments of y are those of a clean trajectory. In Fig. 1 we show $\hat{a}$ and $\hat{b}$ as a function of the relative noise level $\sqrt{\langle \eta_n^2 \rangle / \langle y_n^2 \rangle}$ for equally distributed and for Gaussian noise respectively. The deviation from the correct value $a = 1.4$ and $b = 0.3$ is tremendous for larger noise amplitudes.

The bias due to the "errors-in-variables" problem is not the only way noise enters the estimation of the dynamics from the data. The values of the fit parameters are also affected by statistical fluctuations. How severe this influence is depends on the kind of model one wants to fit.

On our route towards reality, let us now leave the case in which we know the functional form of the equations. In a generic situation we have to approximate F by a model which has to be general enough to reproduce essential features of F. Two distinct approaches have been studied in the literature, local (or piecewise) linear and global nonlinear function fits.

The main idea of locally linear models [Farmer and Sidorowich, 1987; Casdagli, 1989] is to assume that F is at least piecewise differentiable and smooth enough such that the local tangent maps are reasonable approximations to F. The tangent maps themselves are to be estimated using locally linear fits to the data. Typically, one wants to compute the value of $F(x')$ at a point x', knowing a set of values at points x_n, $n = 1, \ldots, N$, and their images. First one forms a small neighborhood $\mathcal{U}(x')$ containing all points $x_{n'}$ close to x'. Within this neighborhood, F is approximated by a linear map

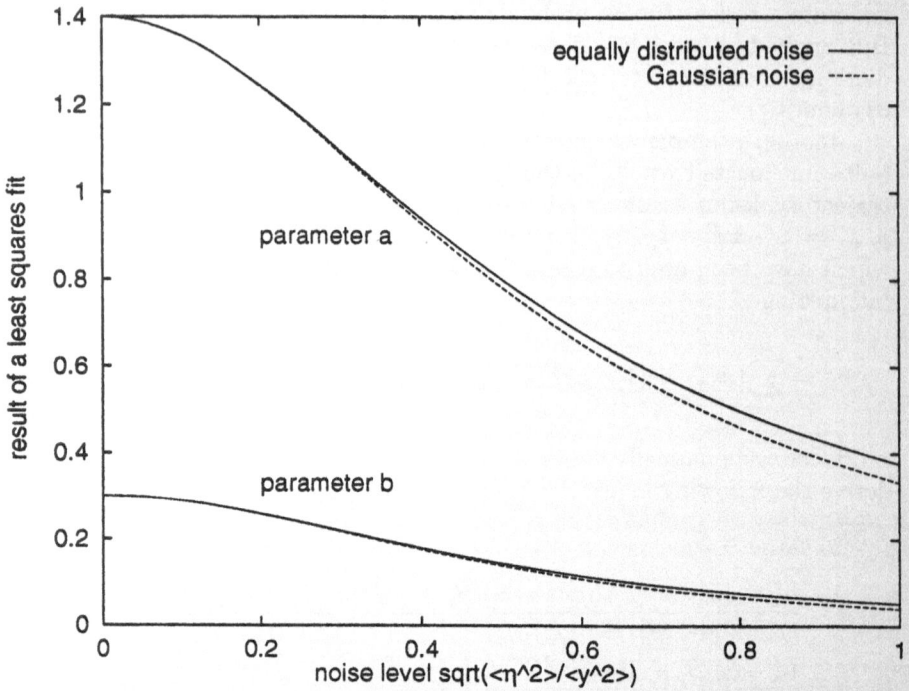

Fig. 1 The analytical results of the least squares fit for the Hénon parameters a and b, Eq. (13). We show the fitted parameters $\hat{a}$ and $\hat{b}$ as a function of the noise level, using trajectories generated with $a = 1.4$, $b = 0.3$.

$$\hat{F}(x) = A'x + b' \ . \tag{15}$$

Once the matrix A' and the vector b' have been determined using a least squares procedure, $\hat{F}(x')$ can be taken as a local estimate for $F(x')$. This fit loses most of its parameters when dealing with delay coordinates. The image of a vector $x_n = (x_n, x_{n-1}, \ldots, x_{n-m+1})$ contains only one new component, $x_{n+1} = (x_{n+1}, x_n, \ldots, x_{n-m+2})$. Thus the matrix A' essentially is a shift operator, apart from the first row, which represents the fact that the new component is, in linear approximation, given by $x_{n+1} = a'x + b$.

This procedure is able to approximate very general functional forms, which makes it very attractive. It has some weak points, though. Obviously, dynamically correct linear fits can only be obtained when the size of the neighborhoods $\mathcal{U}$ is larger than the noise level. Since larger neighborhoods yield poorer linear approximations due to the nonlinearity of F, errors in the dynamics increase with the noise level. Note that increasing the amount of data does not improve the situation as long as no explicit noise reduction step is performed. In the papers of [Abarbanel, 1992 and Brown, Bryant and Abarbanel, 1991] nonlinearity was taken into account by adding a bilinear term to the rhs of Eq. (27), but it is our experience that for noisy data these fits become rather unstable.

Currently, global fits are quite popular [Casdagli 1989; Broomhead and Lowe, 1988; Smith, 1992; Holzfuss and Kadtke, 1993; Gencay and Dechert, 1992] and in many cases quite successful. In delay coordinates, the task is again to fit a scalar function f in an m-dimensional space. Thus the ansatz can in principle be any reasonable set of basis functions; examples include higher order polynomials, radial basis functions, or neural networks. In the first two cases the determination of the parameters still is a linear minimization problem, in the latter case it is nonlinear and therefore more troublesome. Neural nets have the potential of astonishingly good performance, but one must not forget that the theory and application of neural nets is a field of research in itself, often leaving the unexperienced user disappointed.

Since global fits are obtained using all the available data they show less statistical fluctuations than local models, in particular for small data sets. Like local linear fits they react to noise by smoothing out nonlinearities. It is however hard to predict how severe this effect will be for a given global ansatz.

A problem specific for global fits is that the basis may be too small to capture essential features of the surface. A systematic bias is introduced through the choice of basis functions.

4 Noise and Scaling

The complicated dynamics of chaotic systems is generally augmented by a nontrivial geometrical structure in state space. In the last section we discussed what limitations noise poses on predictions. Here we want to focus on the static counterpart of the dynamics, which is given by the probability that a trajectory visits a certain point in space, and by transition probabilities.

We have to require stationarity of the process, which means on the one hand that the governing dynamical laws remain the same as time passes, but on the other hand that the probabilities mentioned above are time invariant. If the map F is dissipative, which is the typical case, trajectories will after some transient time settle down onto an attractor. Only then we can speak of a stationary process, although the governing equations of motions are still the same. On the attractor, the trajectory creates an invariant measure given by the average visitation frequency. An attractor can be a fixed point, i.e., the system remains in a certain state; a limit cycle, corresponding to periodic behavior; or a torus with quasiperiodic dynamics. In these cases the invariant geometric objects are quite familiar.

Much more complicated attractors are possible if the dynamics is chaotic, so called *strange attractors*, which are fractal objects. The latter term means that they exhibit nontrivial structure on all length scales. If we want to

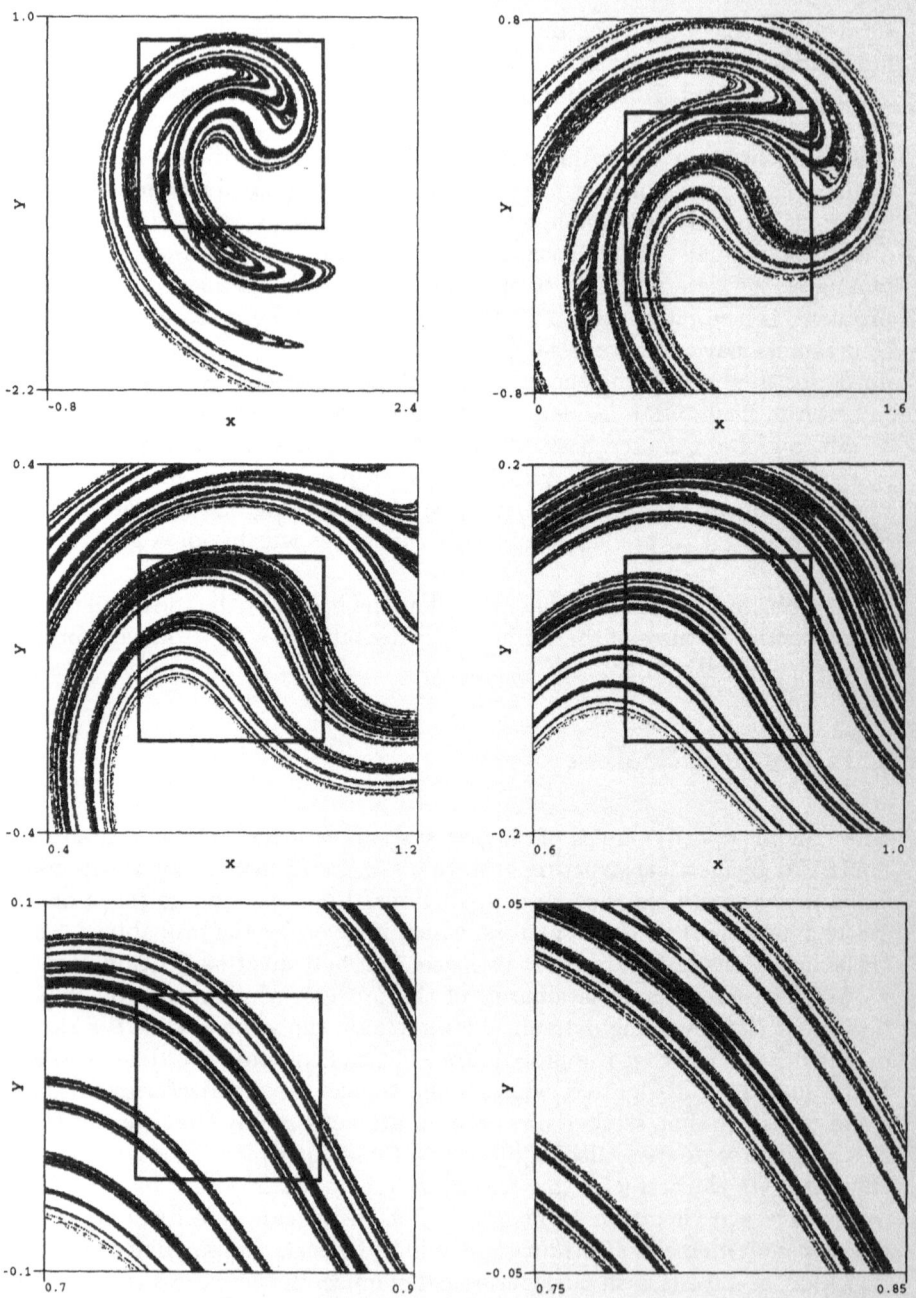

Fig. 2 Successive enlargements of a set of points on the Ikeda attractor [Ikeda, 1979], $z_{n+1} = 1 + 0.9z_n \exp[0.4i - 6i/(1 + |z_n|^2)]$. On smaller length scales, more and more structure becomes visible. Although we do not expect *exact* self-similarity, the line structure repeats itself. Each picture contains about 50000 points.

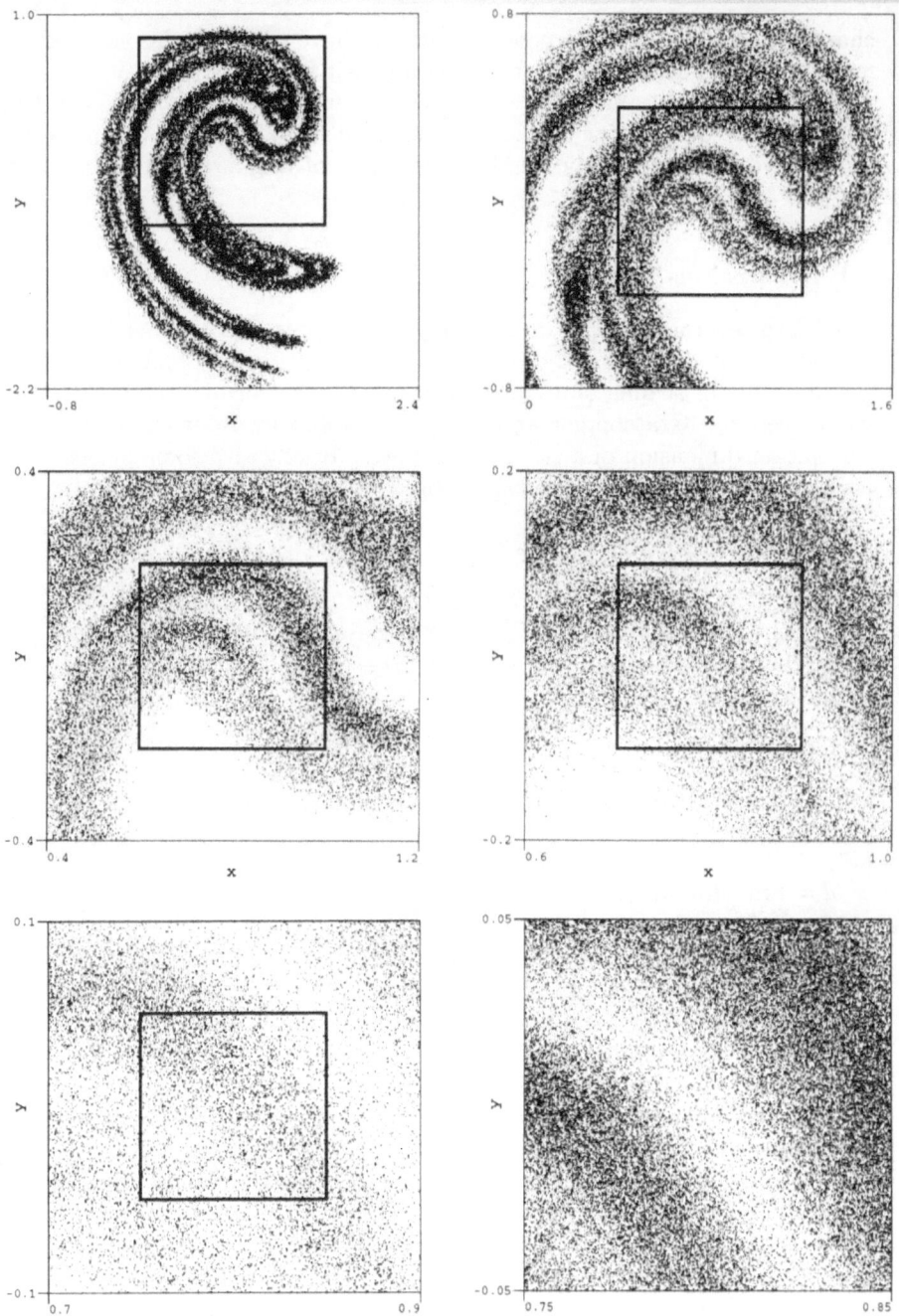

Fig. 3 Same as Fig. 2. Now 2% Gaussian measurement noise has been added to the data. Obviously we can no longer observe the self-similar structure found before.

characterize such a system geometrically we have to study this structure on many length scales, looking for scaling and self-similarity.

Obviously, the range of scales accessible from the data is limited from above by the overall size of the attractor and from below by the noise amplitude. We want to study this limiting effect of measurement noise in this section.

4.1 Dimensions

There are several concepts quantifying scaling properties and self-similarity of a set, most of them use dimension-like quantities. In practice, the most useful probe for scaling of attractor measures is the correlation integral $C(\varepsilon)$ introduced by [Grassberger and Procaccia, 1983] in order to compute the correlation dimension of a strange set. Let us briefly give some properties of estimates of the correlation integral. Let

$$C_N(\varepsilon) = \frac{2}{N(N-1)} \sum_{i,j} \Theta(\varepsilon - \|x_i - x_j\|) , \qquad (16)$$

denote the fraction of pairs of points on the attractor whose distance apart is less than ε. In the limit as $\varepsilon \to 0$ and $N \to \infty$, and in the absence of noise, we have

$$C(\varepsilon) \sim \varepsilon^d , \qquad (17)$$

where d is the correlation dimension of the attractor. Equation (17) can also be written as

$$d = \lim_{\varepsilon \to 0} \lim_{N \to \infty} d(\varepsilon), \qquad d(\varepsilon) = \frac{d}{d \log \varepsilon} \log C_N(\varepsilon) , \qquad (18)$$

which is a more useful definition in practice.

Typically, one reconstructs the attractor in a suitable space of dimension m and computes $C(\varepsilon)$ and its slope $d(\varepsilon)$ as functions of ε. When interested in the correlation dimension d, one would look for a range of ε values where $d(\varepsilon)$ is relatively constant. Here we instead want to study $d_m(\varepsilon)$ as a function of ε and the embedding dimension m to extract information about the noise in the system.

One can distinguish four different types of behavior of $d_m(\varepsilon)$ for different regions of length scales ε (see Fig. 4). For small ε (region I) the lack of data points is the dominant feature. Therefore, the values of $d_m(\varepsilon)$ are subject to large statistical fluctuations. On the other hand, if ε is of the order of the size of the entire attractor (region IV), no scale invariance can be expected.

In between, we can distinguish two regions. Region II is dominated by the noise in the data: the reconstructed points are not restricted to the fractal structure of the attractor but fill the whole phase space available, thus we expect $d_m(\varepsilon) \approx m$, the dimension of phase space. Between regions II and IV

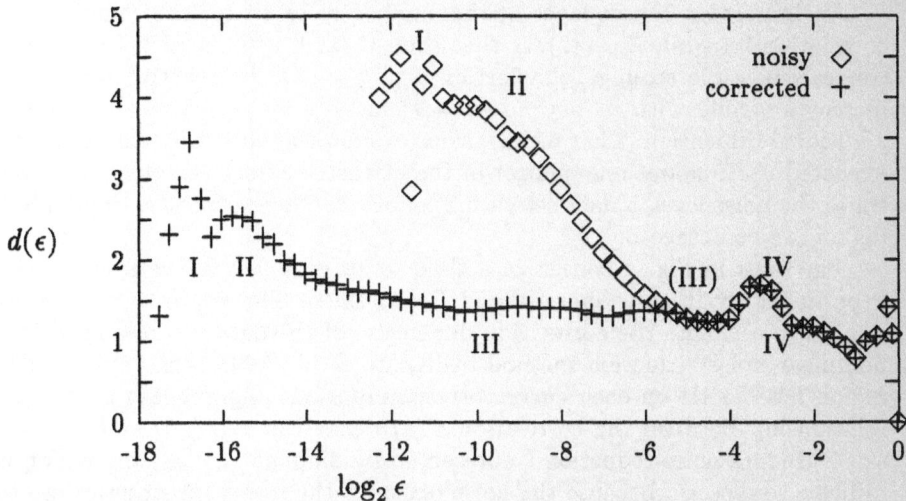

Fig. 4 A typical plot of the correlation dimension as a function of the range of distances ε used to estimate the scaling exponent. Noise obscures the fine-scale structure up to 1/16 of the attractor extent, making an accurate estimate of the attractor dimension impossible. Data from the NMR-laser experiment by [Flepp, Simonet and Brun, 1991], about the noise reduction see [Kantz et al., 1993].

we have region III, where the proper scaling behavior of the attractor may be observed: $d_m(\varepsilon) \sim d$.

For higher values of m, regions II and III will be very distinct: we will see a crossover between $d_m(\varepsilon) \approx m$ below and $d_m(\varepsilon) \approx d$ above the noise level. In [Schreiber, 1993] a formula describing this crossover is derived analytically. Here we only give the result.

Let us consider a signal which can already be faithfully reconstructed in r-dimensional space. If we reconstruct the same signal in $m > r$ dimensions, the effective dimensions $d_m(\varepsilon)$ in m-space are related to those in r-space by the formula

$$d_{m,r}(\varepsilon) \overset{\text{def}}{=} \frac{d_m(\varepsilon) - d_r(\varepsilon)}{m - r}$$

$$= \frac{\varepsilon \exp{-\frac{1}{4\sigma^2}\varepsilon^2}}{\sigma\sqrt{\pi}\, erf(\frac{\varepsilon}{2\varepsilon})} \tag{19}$$

$$= g\left(\frac{\varepsilon}{2\sigma}\right),$$

where we introduced

$$g(z) = \frac{2}{\sqrt{\pi}}\frac{ze^{-z^2}}{\text{erf}(z)}. \tag{20}$$

An important consequence of this analytical result for the shape of the correlation integral, Eq. (19), is that even a small amount of noise already conceals possible scaling behavior: even at $\varepsilon = 3\sigma$ the effective dimension increases visibly with m beyond $m = r$, namely by an amount of 0.2 per additional dimension. That means, since even in the best case scaling can be expected up to about one-quarter of the attractor extent and down to three times the noise level, a data set with 2% noise can give at most a tiny scaling region of two octaves.

The data in Fig. 4 consist of a time series of 40000 values from a laser experiment by [Flepp, Simonet and Brun, 1991]. The curves illustrate attempts to estimate the correlation dimension of the data before and after a nonlinear noise reduction method of [Kantz et al., 1993] was applied. The region labeled III on each curve corresponds to an approximate power law relationship between the correlation $C(\varepsilon)$ defined in Eq. (16) and the ball size ε. Before noise reduction (represented by diamonds), a scaling region is difficult to discern, because the noise obscures the fine scale structure up to 1/16 of the attractor extent. A wide range of estimates of the attractor dimension are possible with the noisy data, depending on the range of ε used. One cannot give any reliable estimate of d for the unprocessed data. Hence, even small levels of noise significantly complicate estimates of the dimension, a quantity that in principle should be straightforward to measure.

4.2 Entropies

While dimensions characterize the scaling properties of the invariant measure which is the probability distribution of the data in state space, entropies do the same for transition probabilities from one part of the state space to another. If the dynamics is regular, all points inside a small ball in the state space will be mapped into another small ball, such that all but one of the transition probabilities are almost zero. Conversely, if we regard a pure noise process, the future of such a set of points is almost undetermined by its present state. Chaotic systems are somewhere in between. Thus entropies measure the loss of information about the state of the system due to the time evolution. However, as in the case of dimensions, a reasonable definition has to involve infinitely small scales and, furthermore, infinite times.

Out of the zoo of entropies we want to select the correlation entropy K_2, since it can be determined numerically quite easily [Grassberger and Procaccia, 1983b]. Again we compute the correlation sum Eq. (16), but this time we do not investigate its dependence on ε but on the embedding dimension m. Then for ideal data one finds

$$C(\varepsilon, m) := \lim_{N \to \infty} C_N(\varepsilon, m) \propto \varepsilon^d \exp(-mK_2) \,, \tag{21}$$

and numerically,

$$K_2(\varepsilon, m) = \ln \frac{C_N(\varepsilon, m)}{C_N(\varepsilon, m+1)} \tag{22}$$

should approximate K_2 for a reasonable range of m and ε. The discussion about the proper range of ε carried out in the last subsection applies here as well. An increase of the scaling exponents $d(\varepsilon)$ with the embedding dimension, as it is typical of region II, corresponds to an overestimation of $K_2(m)$. More precisely, for Gaussian IID noise one finds:

$$K_2(\varepsilon, m) = K_2 - g\left(\frac{\varepsilon}{2\sigma}\right) \ln \varepsilon . \tag{23}$$

In particular, for ε smaller than the noise level the estimate of the entropy is $K_2 - \ln \varepsilon$, which diverges for $\varepsilon \to 0$ as expected. Thus again the variance of the noise has to be sufficiently small compared to the overall size of the attractor, such that there remains a window of ε where $K_2(m)$ becomes independent of ε and thus represents the correct value. Note that noise does not destroy the scaling properties of $C_N(\varepsilon, m)$ in m, but primarily results in an overestimation of K_2.

From the point of view of physics the Kolmogorov–Sinai entropy K_1 is more interesting than the correlation entropy. It differs from K_2 in the way averages are performed; it assigns larger weights to small transition probabilities. This induces numerical problems since small probabilities correspond to rare events. Its computation, which is conveniently performed by a modification of the correlation algorithm [Grassberger and Procaccia, 1984], is much less stable than that of K_2, with respect to a lack of neighbors (small ε), with respect to saturation effects (large ε), and with respect to noise.

The Kolmogorov–Sinai entropy is related to the Lyapunov exponents, which are discussed in the next section. On a generic attractor, K_1 is simply the sum of all positive Lyapunov exponents, such that one can shift the problem to the computation of the latter. In fact, this is the way K_1 usually is computed. Finally we want to mention that K_2 is a lower bound to K_1, such that several consistency checks are possible.

5 Noise and Lyapunov Exponents

An important concept in characterizing chaotic behavior are the Lyapunov exponents. In chaotic systems, the states of two copies of the same system, started with very similar initial conditions, separate on average exponentially with time. The leading contribution to the rate of this separation is given by the largest exponent, the next to leading contribution by the second, etc. These quantities are obtained in a relatively easy way from dynamical equations using information from the dynamics in tangent space, while their estimation from a data series poses some difficulties. This is not the place to review the theory of Lyapunov exponents or methods to obtain them from a time series in detail.[11] We want rather to concentrate on limitations to the current algorithms induced by noise in the data.

5.1 Methods in Real Space

In 1986, Wolf, Swift, Swinney, and Vastano introduced the first method to determine the maximal Lyapunov exponent from a time series. Their method is based on the direct observation of the exponential divergence of nearby trajectories in a reconstructed state space. A close neighbor of the first point of the time series is sought, and the increase of their distance in time is recorded. Eventually, if the dynamics is chaotic, this distance has to become of the order of the attractor size and the expected exponential increase obviously has to cease. Therefore, a new close neighbor for the present point of the trajectory has to be found in approximately the same direction as the old one.

Noise leads to a severe overestimation of the expansion rate. The increase of distances below the noise level is mainly determined by the noise process and not by the deterministic part of the dynamics. For uncorrelated noise, trajectories follow a random walk on small scales, which together with the deterministic exponential divergence yields for the increase of small distances δ with time

$$\delta(n) \approx \alpha\sigma\sqrt{n} + \beta e^{\lambda_+ n} . \tag{24}$$

Thus for small n one would find a diverging effective exponent, which would lead to an estimate of the Lyapunov exponent which is substantially too large. In [Wolf et al., 1986] it was suggested that only neighbors with an initial separation larger than the noise level be used, which means that the procedure of finding a new appropriate neighbor has to be performed much more frequently. As soon as the distance becomes of the order of the attractor size, due to saturation and folding processes it will no longer increase with an exponential law, such that the maximum distance should be of the order of $1/5$ (assuming normalized data). If the maximal Lyapunov exponent of the clean dynamics is λ_+, on average the distance will grow by a factor of e^{λ_+} in each time step. If we make the optimistic assumption that the diffusive regime does not exceed twice the noise level, the maximal acceptable noise amplitude is

$$2\sigma < 1/5e^{-\lambda_+ n} , \tag{25}$$

if we require n time steps between successive searches for new neighbors. Obviously, noise must not be larger than $\sigma_{max} = 1/10e^{-\lambda_+}$. To obtain a

---

[11]The interested reader is referred to the original literature [Benettin et al., 1980; Eckmann et al., 1986, Sano and Sawada, 1985; Wolf et al., 1985] and to recent treatments of this topic by [Brown, Bryant and Abarbanel, 1991; Gencay and Dechert, 1992; Parlitz, 1992; Rosenstein, Collins, and De Luca, 1993; Kantz, 1993].

reasonable estimate of λ_+ for the Hénon map one would require, say $n > 5$ to make sure we do not leave the unstable manifold. With $\lambda_+ = 0.4169$ this means $\sigma_{\max} < 0.012$, i.e. less than 1.2%.

Several variations and improvements of the Wolf algorithm have been suggested, which we do not want to review here, since they all suffer from the same shortcoming: these algorithms always yield only a bare number, the calculated Lyapunov exponent, without any way to estimate the error other than by varying the parameters of the algorithm. Also due to their numerical instability, the Wolf algorithm and its variants are not the methods of choice.

Nevertheless, the idea of an algorithm relying directly on the unprocessed data is quite charming. Recently, a considerable breakthrough was obtained independently in [Rosenstein, Collins and De Luca, 1993] and [Kantz, 1993]. Let us give a very short description following the realization by [Kantz, 1993] which suffers less from statistical errors and yields a larger scaling range. The main idea is not to absorb the time dependence of the distances in a single number but to record it explicitly, such that one is able to select the appropriate length scale and range of times from the output instead of setting these parameters in advance. To determine the Lyapunov exponents, one has to compute

$$
S(\Delta n) = \frac{1}{N-m} \sum_{n=m+1}^{N} \log \left(\frac{1}{|\mathcal{U}_n|} \sum_{x_{n'} \in \mathcal{U}_n} |x_{n'+\Delta n} - x_{n+\Delta n}| \right) \tag{26}
$$

for different embedding dimensions m and neighborhoods of different sizes ε. Here, $x_n = (x_n, x_{n-1}, \ldots, x_{n-m+1})$ is a delay vector of dimension m, and $\mathcal{U}_n$ is the set of all other delay vectors in an ε-neighborhood of x_n. $|\mathcal{U}_n|$ is the number of elements in $\mathcal{U}_n$. If $S(\Delta n)$ as a function of Δn shows a linear regime, the slope can be interpreted as the maximal Lyapunov exponent. This is illustrated in Fig. 5 for a normalized Hénon time series with 1% measurement noise: Only on scales between the noise level, $\log \varepsilon_{\min} \approx -5.33$ and $\varepsilon_{\max} \approx e^{-2} = 0.135$ can the expected linear behavior be found, yielding an average slope of all the curves shown of 0.41 ± 0.015, which is a very good estimate of the maximal Lyapunov exponent $\lambda_+ = 0.4169 \pm 0.0001$.

Furthermore, Fig. 5 shows impressively that neighbors closer than the noise level yield very large effective exponents (regard the two lower bundles of curves). Due to the averaging of multiple curves, scaling extends down to the noise level resulting in an admissible noise amplitude *twice* as large as the estimate for the Wolf method, Eq. (25), i.e. 2.4% for the Hénon example.

5.2 Exploiting the Dynamics in Tangent Space

Another class of methods obtains the momentary expansion rates from the dynamics in tangent space, in close analogy to the methods for known dynamical equations [Benettin et al., 1980]. This approach yields not only the

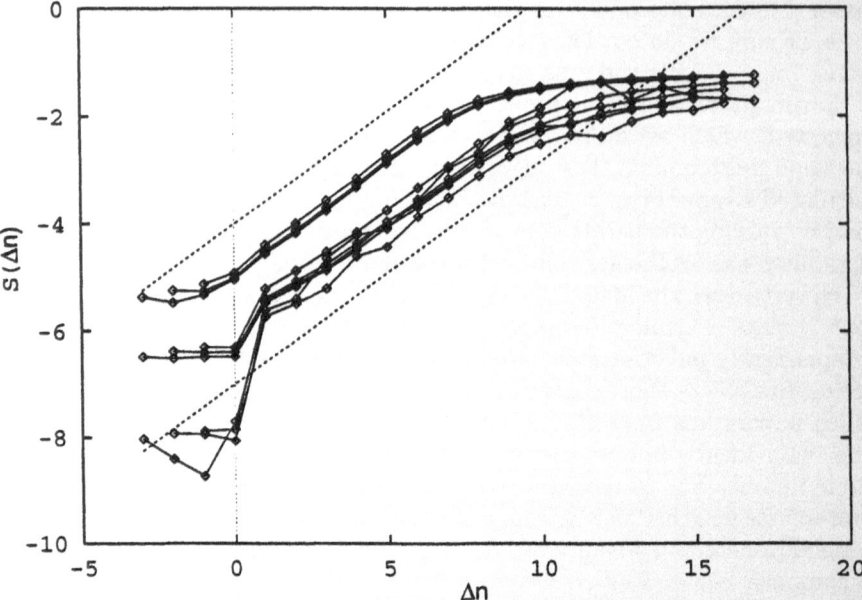

Fig. 5 The function $S(\Delta n)$, Eq. (26), for a Hénon time series of length 5000 with 1% additive noise. The three bundles of curves correspond to neighborhood sizes $\varepsilon = 0.001$, 0.004 and 0.016 from bottom to top, and each bundle consists of curves with embedding dimension from 2 to 4. The coordinates used to determine the neighborhoods are denoted by $\Delta n = 0, \ldots, (1 - m)$. The slopes of the two dashed lines are the accurate values for λ_+.

largest Lyapunov exponent, but m exponents for an m-dimensional embedding. Since the dimensionality of the underlying dynamical system generally is unknown, many of the negative exponents may be spurious. For a discussion of this problem see [Eckmann et al., 1986; Gencay and Dechert, 1992; Parlitz, 1992].

These algorithms can cope with much higher noise levels than the real-space algorithms. However, they too provide no error estimate. In particular, it is hard to know whether the numbers describe some true exponential behavior or whether they are artifacts of some other functional dependence.

The Lyapunov exponents are determined by the logarithms of the eigenvalues of the product of the Jacobians of F along the trajectory. Thus one has to know the local Jacobians with sufficient accuracy. This problem is closely related to the problem of fitting the dynamics for predictions, which has been discussed above.

The original works, [Sano and Sawada, 1985] and [Eckmann et al., 1986], perform linear fits of the local dynamics in small volumes of phase space. This process immediately yields the corresponding local Jacobian:

$$x_{n+1} = A^{(n)} x_n + b^{(n)}, \quad J^{(n)} = A^{(n)}. \tag{27}$$

The matrix $A^{(n)}$ and vector $b^{(n)}$ are obtained by a least squares fit in a suitable neighborhood around $(x_n,\ x_{n+1})$. As also discussed in connection with prediction, this kind of fitting the dynamics has the considerable advantage that it incorporates minimal prior information about the shape of F, except for its smoothness. However, it can suffer from statistical errors if only small amounts of data are available. Moreover, the noise present in the data requires the choice of sufficiently large neighborhoods in order to determine J correctly. This may lead to systematic errors due to the curvature of the attractor.

Fitting instead a nonlinear global ansatz for F reduces statistical problems for short data sets substantially. (We refer to the section on prediction for more material on local vs global fits.)

No matter what kind of model is used, the resulting approximation is subject to the bias induced by the errors in the independent variables. Glancing back at Eq. (14) and Fig. 1, it is obvious that for large noise levels the Lyapunov exponents computed by the Jacobians of this fitted map are systematically wrong. This is shown in Fig. 6. The two broken curves represent the positive and negative Lyapunov exponent obtained by the local Jacobians with the parameters from Fig. 1 on a noisy trajectory of length 10000 with the indicated noise level. For comparison, the continuous lines are the corresponding values obtained with the correct Jacobians on the same noisy trajectories. One clearly observes that the additive noise on the trajectory is relatively harmless as long as the correct dynamics is known.

Some authors [Farmer and Sidorowich, 1988] suggested that one fit the dynamics underlying a given noisy time series and generate a new trajectory by iteration of the fitted dynamics. This "bootstrapping" yields a clean trajectory of arbitrary length of this model system, which can be used for further analysis. The dotted curves in Fig. 6 show the resulting exponents of such a procedure. Obviously, for large noise levels the fitted dynamics creates a completely different attractor, which is no longer compatible with the original noisy time series. Note that for more than 30% noise the surrogate attractor is a periodic orbit.

Thus we have shown that one can recover the correct Lyapunov exponents with good accuracy even for high noise levels if one has a faithful estimate of the underlying dynamics. The latter could be obtained by an unbiased fit, to which problem a forthcoming paper will be devoted. An alternative solution is to apply nonlinear noise reduction and afterwards perform a usual fit. The success of such a procedure [Grassberger et al., 1993] is shown by the diamonds in Fig. 6, which represent the Lyapunov exponents obtained from the noisy data sets after noise reduction.

To finish this section we want to recall the Kaplan–Yorke conjecture, which states that

$$D_{\mathrm{KY}} = l + \frac{\sum_{i=1}^{l} \lambda_i}{|\lambda_{l+1}|},$$

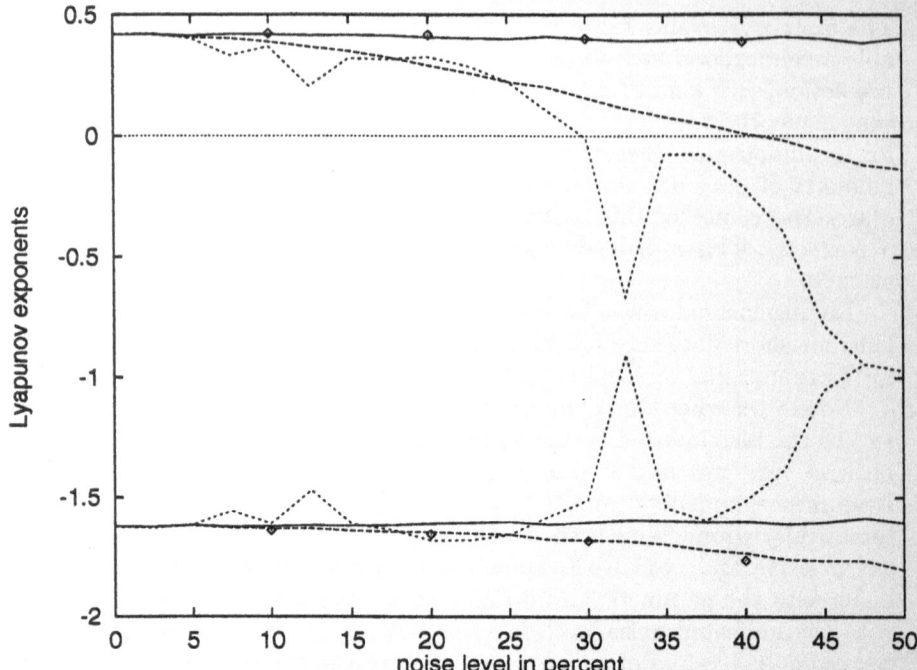

Fig. 6 Lyapunov exponents obtained from the local Jacobians for noisy Hénon trajectories. The continuous curves show the values obtained from the exact Jacobians along the noisy trajectories. For the dashed curves, the fitted parameters from Fig. 1 were used. The dotted curves are obtained from surrogate trajectories generated by the fitted dynamics. The diamonds represent the results when a noise reduced trajectory is used to fit the parameters but the Jacobians are taken from the original data.

where l is chosen such that $\sum_{i=1}^{l} \lambda_i \geq 0$ and $\sum_{i=1}^{l+1} \lambda_i < 0$ is equal to the information dimension D_1 of the attractor. Since D_2 is a lower bound of the latter, one should check for consistency whether $D_{KY} \geq D_2$.

6 Nonlinear Noise Reduction

When thinking about noise reduction, Fourier based filtering techniques come to mind. For data from mainly linear sources they are quite successful, since the Fourier spectrum often allows clear distinction between signal and noise. However, since data from nonlinear sources generally possess a broad band spectrum themselves, such a distinction fails. Only for chaotic flows recorded with a very small sampling time (compared to the internal time scale) may a low pass filter be applicable. Unfortunately, certain IIR-filters used as real-time filters in experiments may change the dimension of the systems attractor [Badii et al., 1988].

Within nonlinear time series analysis several methods for nonlinear noise reduction have been suggested. They all make use of the fact that data points which follow one another in time are dynamically correlated and that, due to the deterministic features, points in the same region of the state space have to behave similarly. Methods which are most apt for scalar time series without any a priori knowledge of the dynamics make use of projections onto a reconstructed local manifold containing the attractor [Schreiber and Grassberger, 1991; Cawley and Hsu, 1992; Sauer, 1992; Grassberger et al., 1993]. It seems that nonlinear noise reduction is currently not very much in use, although it has been shown using experimental data that these algorithms can be applied with a reasonable effort and that they yield stable results [Kantz et al., 1993]. Furthermore, for trajectories with high noise levels a very simple but robust variant of nonlinear noise reduction is applicable [Schreiber, 1993]. We only want to encourage the reader to try one of these algorithms. For this purpose we have shown in Fig. 4 the improvement of dimension estimates obtained by applying the method of [Grassberger et al., 1993] to experimental data, and in Fig. 6 we show the improved Lyapunov exponents.

7 Conclusions and Prospects

First the bad news: Noise degrades the whole field of nonlinear time series analysis so strongly that many algorithms are rendered useless even by a few percent of noise. Since many nonlinear phenomena are observed in non-laboratory systems (in the atmosphere, ocean, human body, economics) and thus the dynamical noise cannot be controlled, one might think that the most interesting data cannot be processed. However, we do not adopt this pessimistic attitude.

Fortunately, there is good news as well: First, the influence of noise in many instances is well understood. This helps to identify noise, to estimate its strength and to see how much of the deterministic structure is still visible. Second, in many situations nonlinear noise reduction works satisfactorily. Many experiments yield data which are on the verge of being processable by nonlinear tools, and after nonlinear noise reduction one can successfully treat them. Third, there are still some things that can be done even in the presence of strong noise. Global fits are statistically quite stable against noise. The errors-in-variables problem seems to be tractable, such that we hope to perform unbiased fits of the dynamics. This having been done, one could in fact produce a new noise-free trajectory. A test of whether these surrogate data are compatible with the original noisy ones exists [Kantz, 1994], such that erroneous results like those presented in Fig. 6 for the naive "bootstrapping" can be ruled out. Thus work is in progress which makes us optimistic about the problem of noise in data.

There are other hard problems which remain, for example, too short observation times, non-stationarity, and the problem of spatio-temporal chaos.

In the last case, data do not represent a low dimensional attractor, but rather a process where characteristic quantities like dimensions and Lyapunov exponents become intensive quantities. How can we reconstruct the corresponding state space? Although some theoretical results are already available, they are far from being applicable to time series. Thus for many realistic systems we are only on the second step of a long staircase towards satisfactory data analysis with predictive power.

References

Badii, R., Broggi, G., Derighetti, B., Ravani, M., Ciliberto, S., Politi, A., and Rubio, M.A. (1988): Phys. Rev. Lett. **60**, 979.

Benettin, G., Galgani, L., Giorgilli, A., and Strelcyn, J.-M. (1980): "Lyapunov characteristic exponents for smooth dynamical systems and for Hamiltonian systems: A method for computing all of them". Meccanica **15**, 9.

Broomhead, D. and Lowe, D. (1988): "Multivariable function interpolation and adaptive networks". Complex Systems **2**, 321.

Brown, R., Bryant, P., and Abarbanel, H.D.I. (1991): "Computing the Lyapunov spectrum of a dynamical system from an observed time series". Phys. Rev. A **43**, 2787.

Casdagli, M. (1989): "Nonlinear prediction of chaotic time series". Physica **35D**, 335.

Casdagli, M., Eubank, S., Farmer, J.D., and Gibson, J. (1991): "State space reconstruction in the presence of noise". Physica **51D**, 52.

Casdagli, M., Eubank, S. (Eds.) (1992): "Nonlinear Modelling and Forecasting". *Santa Fe Studies in the Science of Complexity. Proc.* Vol. XII, Reading, MA.

Cawley, R. and Hsu, G.-H. (1992): "Local-geometric-projection method for noise reduction in chaotic maps and flows". Phys. Rev. A **46**, 3057.

Eckmann, J.P. and Ruelle, D. (1986): "Ergodic theory of chaos and strange attractors". Rev. Mod. Phys. **57**, 617.

Eckmann, J.-P., Kamphorst S.O., Ruelle, D., and Ciliberto, S. (1986): "Lyapunov exponents from a time series". Phys. Rev. A **34**, 4971.

Farmer, J.D. and Sidorowich, J.J. (1987): Predicting chaotic time series, Phys. Rev. Lett. **59**, 845.

Farmer, J.D. and Sidorowich, J.J. (1988): "Exploiting chaos to predict the future and reduce noise". In Y. C. Lee, Ed., *Evolution, Learning and Cognition.* World Scientific, Singapore.

Finardi, M., Flepp, L., Parisi, J., Holzner, R., Badii, R., and Brun, E. (1992): Phys. Rev. Lett. **68**, 2989.

Flepp, L., Holzner, R., Brun, E., Finardi, M., and Badii, R. (1991): Phys. Rev. Lett. **67**, 2244;

Gencay, R. and Dechert, W.D. (1992): "An algorithm for the n Lyapunov exponents of an n–dimensional unknown dynamical system". Physica **59D**, 142.

Grassberger, P. and Procaccia, I. (1983a): "Characterization of strange attractors". Phys. Rev. Lett. **50**, 346.

Grassberger, P. and Procaccia, I. (1983b): "Estimation of the Kolmogorov entropy from a chaotic signal". Phys. Rev. A **28**, 2591.

Grassberger, P. and Procaccia, I. (1984): "Dimensions and entropies from a fluctuating dynamics approach". Physica **13D**, 34.

Grassberger, P., Schreiber, T., and Schaffrath, C. (1991): "Nonlinear time-sequence analysis". Int. J. of Bifurcation and Chaos 1, 521.

Grassberger, P., Hegger, R., Kantz, H., Schaffrath, C., and Schreiber, T. (1993): "On Noise Reduction Methods for Chaotic Data". Chaos 3, 127.

Holzfuss, J. and Kadtke, J.B. (1993): "Global nonlinear noise reduction using radial basis functions". Int. J. Bifurcation and Chaos 3, 589.

Ikeda, K. (1979): Opt. Commun. 30, 257.

Kantz, H. and Grassberger, P. (1985): "Repellers and long-lived chaotic transients". Physica 17D, 75.

Kantz, H. (1993): "A robust method to estimate the maximal Lyapunov exponent of a time series". Wuppertal preprint.

Kantz, H., Schreiber, T., Hoffmann, I., Buzug, T., Pfister, G., Flepp, L.G., Simonet, J., Badii, R., and Brun, E. (1993): "Nonlinear noise reduction: A case study on experimental data". Phys. Rev. E 48, 1529.

Kantz, H. (1994): "Quantifying the distance between fractal measures". Wuppertal preprint.

Kostelich, E.J. (1992): "Problems in estimating dynamics from data", Physica 58D, 138.

Lorenz, E.N. (1963): "Deterministic nonperiodic flow". J. Atmos. Sci. 20, 130.

Parlitz, U. (1992): "Identification of true and spurious Lyapunov exponents from time series". Int. J. of Bifurcation and Chaos 2, 155.

Rosenstein, M.T., Collins, J.J., and De Luca, C.J. (1993): "A practical method for calculating largest Lyapunov exponents from small data sets". Physica 65D, 257.

Sano, M. and Sawada, Y., (1985): "Measurement of the Lyapunov spectrum from a chaotic time series". Phys. Rev. Lett. 55, 1082.

Sauer, T., Yorke, J.A., and Casdagli, M. (1991): "Embedology". J. Stat. Phys. 65, 579.

Sauer, T. (1992): "A noise reduction method for signals from nonlinear systems". Physica 58D, 193.

Schreiber, T. and Grassberger, P. (1991): Phys. Lett. A 160, 411.

Schreiber, T. (1993a): "Extremely simple nonlinear noise reduction method". Phys. Rev. E 47, 2401.

Schreiber, T. (1993b): "Determination of the noise level of chaotic time series". Phys. Rev. E 48, R13.

Smith, L.A. (1992): "Identification and prediction of low-dimensional dynamics". Physica 58D, 50.

Wolf, A., Swift, J.B., Swinney, H.L., and Vastano, J.A. (1985): "Determining Lyapunov exponents from a time series". Physica 16D, 285.

A Discriminant Procedure for the Solution of Inverse Problems for Nonstationary Systems

Oleg L. Anosov and Oleg Ya. Butkovskii

Abstract

The inverse problem of nonlinear dynamics is analyzed, with reference to nonstationary chaotic systems. Numerical procedures are developed for the reconstruction of differential equations directly from experimental series. To consider the procedure of model identification in state space, a differential identification scheme along two time intervals (windows) has been introduced, typical of discriminant analysis. Such a scheme is shown to reliably detect nonstationarities caused by changes both in control parameters of the system itself and external forces. High sensitivity of the differential schemes to the parameter variation is also exhibited in the specific case while determining the model type. The efficiency of such a procedure is demonstrated using the examples of a noisy discrete map and the Rossler chaotic system with step-wise and sinusoidally varying changes in control parameter. The differential procedure suggested is capable of revealing abrupt control parameter changes as small as 0.5%, whereas standard procedures of statistical discrimination such as average value and variance show no any meaningful changes in similar conditions.

1 Introduction

The problem of predicting the evolution of a physical system from a data set (i.e. a "time series") is ubiquitous in the natural sciences. In this work we will try to outline some approaches to the problem of such dynamical forecasting, i.e., forecasting based on certain dynamical laws, and to the problem of revealing both small abrupt changes in parameters, and those due to external driving.

If a process is judged to be "random" it is only natural to turn to the statistical methods of prediction originally worked out by N. Wiener. But if there are signs or suspicions that the process under study has a dynamical origin, it is more expedient to try a dynamical forecast based on relevant regularities. Fundamentals for dynamical prediction were developed as a result of long studies, initiated by A.N. Kolmogorov and D. Gabor [see, for instance, Ljung, 1987]. Attempts to apply a dynamical approach to the analysis of chaotic time series are reported in [Breeden and Hübler, 1990; Gouesbet, 1991; Brush and Kadtke, 1992; Gribkov et al., 1994a]. The difference between the statistical and dynamical forecasts can be truly significant.

The statistical forecast starting from probabilistic and correlation structure is usually limited by relatively short intervals of time comparable with the autocorrelation time τ_c of the process. For systems known to have a dynamical origin, it seems logical to use dynamical forecasting. By way of example, we can refer to weather forecasting in which dynamical methods based on hydrodynamical equations for air flows are used to advantage [Mason and Treas, 1986]. Admittedly, the dynamical methods also fail to predict a process's behavior far into the future. One of the main reasons for this is the local instability inherent in most of the phenomena around us and governing such processes as a dynamical chaos.

General physical analysis of predictability limits was carried out in [Kravtsov, 1989] and in a recent collective monograph [Kravtsov (Ed.), 1993], see especially the paper by Kravtsov (pp. 173–203) of that book. Some illustrations of general principles and practical algorithms for solving inverse problems in nonlinear dynamics are given in [Gribkov et al., 1994b; Aivazyan, et al., 1989].

A peculiarity of the equations giving dynamical chaos is that they fail to provide reliable long-term forecasts. The point is that errors in initial conditions, along with other hindering factors (noise inevitable in real systems, small inaccuracies occurring in model equations, a slight nonstationarity of a system etc.), tend to grow exponentially under local instability conditions, so that after some finite time all the worker's efforts are in vain.

Taking A to be the typical magnitude of oscillations under study and δ the perturbation in the system occurring due to one or another factor, the limit time of predictable behavior $\tau_{\lim}$ can be written as:

$$\tau_{\lim} = \frac{1}{2\lambda_+} \ln \frac{A}{\delta} , \tag{1}$$

where λ_+ is the maximum Lyapunov exponent giving the perturbation growth rate.

Although the prediction time is limited, as follows from (1), it can exceed (and often does exceed!) the autocorrelation time τ_c. It is the ratio $\tau_{\lim}/\tau_c$ that characterizes the gain provided by the dynamical forecast as against the statistical forecast. The main question is how to realize this gain when the dynamical equations governing a system's behavior are not known a priori. To reconstruct these equations by analyzing a time series is the "inverse problem" of nonlinear dynamics.

2 Discriminant Procedure for Identification of Nonstationarities in Nonlinear Processes

Procedures for reconstructing a nonstationarity in an observed process $x(t)$ are identical to "discriminating" the characteristics of $x(t)$, measured in two adjacent time intervals, say, in time-windows $[t_1, \ t_1 + T_1]$ and $[t_2, \ t_2 + T_2]$, where t_2 usually equals $t_1 + T_1$ and time intervals T_1 and T_2 are usually equal to each other.

The distinction between characteristics of an observed process $x(t)$ in two neighboring intervals may be detected by the analysis of a "discriminant function" $d[x(t), \ a]$, where $a = (a_1, \ a_2, \ldots, \ a_m)$ is a parameter set of a system under consideration.

Standard discriminant analysis deals with the statistical characteristics of the process $x(t)$ under study and treats the discriminant function values d_1 and d_2 as moving averages of x, or as variances of x inside time-windows 1 and 2 correspondingly. Unfortunately, statistical analysis is practically useless when the system under consideration manifests chaotic behavior. Statistical characteristics of the process $x(t)$ change only insignificantly even if the system parameters change noticeably. That is why we suggest using a discriminant procedure (comparison of discriminant function values d_1 and d_2 in two neighboring windows), but for a quite different type of discriminant function.

We consider it reasonable to use for d a nonlinear function of $x(t)$ and its derivatives, which turns out to be zero when $x(t)$ is an exact solution of the nonlinear equations describing the system under consideration in a noiseless situation. In other words, the equation $d[x(t), \ a] = 0$ is simply a form of the governing equation, which the system obeys hypothetically. The equation of this type contains indefinite parameters a_k, which must be determined. In this case, the time behavior of the coefficient set $a = (a_1, \ a_2, \ldots, \ a_m)$ testifies to nonstationary properties of the system.

There is another approach which uses the Fisher criterion

$$H = \frac{[M_1(t) - M_2(t)]^2}{[S_1^2(t) + S_2^2(t)]} \ . \tag{2}$$

Here $M_1(t)$ and $M_2(t)$ are average values of a discriminant function $d[x(t), \ a]$ within two adjacent windows, and $S_1^2(t)$ and $S_2^2(t)$ are the corresponding variations of $d[x(t), \ a]$. The new point is that the nonlinear discriminant function $d[x(t), \ a]$ described above enters in the Fisher criterion instead of statistical characteristics of the process $x(t)$ itself.

One may expect that the faster parameters change the larger criterion H. That is why we have applied such a procedure to revealing firstly a stepwise change in parameters of nonlinear system. Preliminary results of the discriminant analysis of nonstationary nonlinear time series were presented in [Anosov et al., 1995].

2.1 Nonstationary Discrete Systems (Logistic Map)

Let a sequence $x(n)$ obey a nonlinear map

$$x(n+1) = F[x(n), \, b] \, , \tag{3}$$

where $b = \{b_1, \, b_2, \ldots, \, b_m\}$ is a system parameter set. The integer number n plays the role of discrete time here. As a model discriminant function we choose the difference between left and right sides of (3):

$$d[x(n), \, a] = x(n+1) - F[x(n), \, a] \tag{4}$$

where a is a set of so-far indefinite parameters.

The discriminant function d (4), turns out to be zero in ideal, noiseless conditions, where the parameters a_k to be fitted coincide with the parameters b_k of the real map (3): $a_k = b_k$. The parameters a_k and b_k may be thought of as coefficients of a polynomial, approximating a nonlinear map:

$$F[x(n), \, a] \approx a_0 + a_1 x(n) + a_2 x^2(n) + \ldots + a_m x^m(n) \, .$$

When the parameter b_k of the real system changes abruptly, the values $x(n)$ obey different equations (3) before and after the moment of parameter change. The resulting change in governing equation (3) may be detected by analysis of the H criterion, Eq. (2). The specific algorithm for the detection of moment of parameter change implies not minimization of the discriminant function as above, but maximization of the Fisher criterion by a proper choice of parameters a_k. The efficiency of the algorithm suggested is demonstrated by the example of the logistic map

$$x(n+1) = r(n)x(n)[1 - x(n)] \, , \tag{5}$$

where $r(n)$ is a time-varying control parameter. Figure 1(a) shows a sequence $x(n)$ generated by a logistic map (5) with varying parameter $r(n)$. We suppose that the control parameter changes abruptly at the moment $n = n_a$ from the initial value $r_a = 3.80$ to final value $r_b = 3.82$. At the moment $n = n_b$, the control parameter $r(n)$ returns to the initial value $r = r_a$. The increment of r is as small as 0.02. Thus $r(n)$ changes according to

$$r(n) = \begin{cases} r_a, & \text{when } n \leq n_a, \\ r_b, & \text{when } n_a \leq n < n_a, \\ r_a, & \text{when } n \geq n_b, \end{cases}$$

shown in Fig. 1(b). Both values r_a and r_b exceed the critical value $r_{cr} = 3.5966$ corresponding to the onset of chaos, so the sequence $x(n)$ at Fig. 1(a) is of chaotic nature. Figure 1(a) does not manifest any visible change in behavior.

Meanwhile the discriminant analysis is capable of revealing abrupt change in parameters. Let us consider the quadratic discriminant function

$$d[y(n), \, a] = y(n+1) - a_1 y(n) - a_2 y^2(n) \, , \tag{6}$$

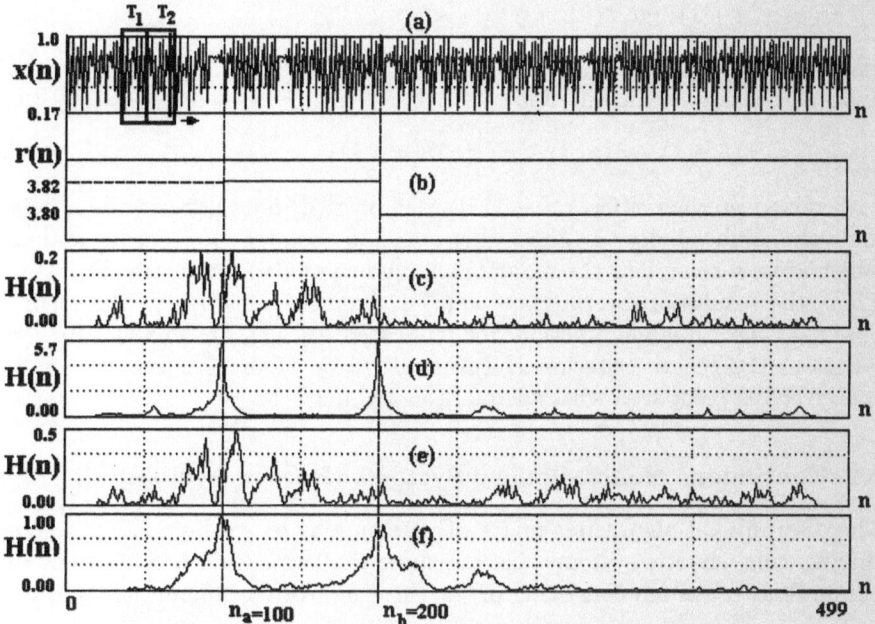

Fig. 1 (a) The sequence $x(n)$, generated by the logistic map with step-wise change in the control parameter $r(n)$ (b). The Fisher criterion $H(n)$ with adequate nonlinear discriminant function (6) shows sharp peaks at the moments of abrupt change of parameter r (d). Adding 0.3% noise makes the peaks wider and lower (f). Inadequate discriminant functions (truncated, Eq. (7), and "mixed", Eq. (8)) provide much lower values of $H(n)$, (c) and (e), respectively.

Table 1

Interval	$n = [0, \, n_a]$	$n = [n_a, \, n_b]$	$n = [n_b, \, 499]$
a_1	-3.797 ± 0.011	-3.817 ± 0.042	-3.801 ± 0.006
a_2	$+3.798 \pm 0.008$	$+3.817 \pm 0.033$	$+3.801 \pm 0.005$

which coincides with the logistic map (5) when $a_1 = -a_2 = r$. Results of the Fisher criterion computation with windows length $T_1 = T_2 = 20$ are shown in Fig. 1(d). The Fisher criterion demonstrates obvious peaks approximately as high as 5.7, though the increment of control parameter was as small as 0.5% (!). The identified parameter values are given in Table 1.

When 0.3% noise was added the peaks became wider and lower ($H_{\max} \approx 1.0$), as may be seen from Fig. 1(f).

Discriminant functions of other types give much worse results. Figure 1(c) contains a plot of Fisher criterion for truncated (linear) discriminant function

$$d_{\text{tr}} = x(n+1) - a_1 x(n) \; . \tag{7}$$

The maximal value $H_{\max}$ happens to be as small as 0.2. The "mixed" quadratic discriminant function

$$d_{\text{mix}} = x(n+1) - a_1 x(n) - a_2 x(n) x(n+1) \tag{8}$$

with mixed product $x(n) \cdot x(n+1)$ instead of $x^2(n)$ does not provide a maximal value higher than 0.5. The Fisher criterion acquires spontaneous maxima which are not related to the moments n_a and n_b of $r(n)$ jumps (see Fig. 1(e)). Thus, the results of discriminant analysis are very sensitive to the form of nonlinear discriminant function: for truncated Eq. (7), or "mixed", Eq. (8), functions the Fisher criterion H turns out to be 25 or 10 times smaller respectively as compared with adequate model map (6).

2.2 Continuous Nonstationary Systems (Rossler System)

The discriminant algorithms under discussion may be readily applied to continuous time chaotic systems. Let's assume that we obtain a continuous dynamic Rossler system consisting of the three first-order differential equations

$$\dot{X} = -Y - Z \; ; \quad \dot{Y} = X + aY \; ;$$

$$\dot{Z} = b + rZ + XZ \; . \tag{9}$$

Let us form three discriminant functions as a difference between left and right parts of each system of equations (9).

$$d_x(\dot{X}, Y, Z; \alpha) = \dot{X} + \alpha_1 Y + \alpha_2 Z \; ,$$

$$d_y(\dot{Y}, X, Y; \beta) = \dot{Y} - \beta_1 X - \beta_2 Y \; , \tag{10}$$

$$d_z(\dot{Z}, X, Z; \gamma) = \dot{Z} - \gamma_0 - \gamma_1 Z - \gamma_2 XZ \; .$$

As in the previous example, each of the discriminant functions (10) in the case of ideal, noiseless conditions will tend to zero if its parameters are equal to those of the corresponding equation for the Rossler system, Eq. (9):

$$\begin{aligned}
d_x(\dot{X}, Y, Z; \alpha) &\to 0 && \text{when} && \alpha_1 = \alpha_2 = 1 \; , \\
d_y(\dot{Y}, X, Y; \beta) &\to 0 && \text{when} && \beta_1 = 1 \, , \beta_2 = a \; , \\
d_z(\dot{Z}, X, Z; \gamma) &\to 0 && \text{when} && \gamma_0 = b \, , \gamma_1 = r \, , \gamma_2 = 1 \; .
\end{aligned} \tag{11}$$

This permits us, using the corresponding discriminant functions of Eq. (10), to find the values of the parameters a, b, r for the Rossler system.

In the given example we used values in Eq. (9) equal to $a = b = 0.2$ and $r = -4.60$, which corresponds to the chaotic conditions of the Rossler system. The numerical solution of the resulting system

$$\dot{X} = -Y - Z \; ; \quad \dot{Y} = X + 0.2Y \; ;$$

$$\dot{Z} = 0.2 - 4.6Z + XZ \;, \tag{12}$$

allowed us to form the data set with the time series values X, Y, Z-variables and their first derivatives. Adjusting the discriminant procedure with the time windows $T_1 = T_2 = 20$ to this file made it possible to restore the system equations as follows

$$\dot{X} = -1.007Y - 1.015Z \;; \quad \dot{Y} = 0.992X + 0.183Y \;;$$

$$\dot{Z} = 0.253 - 4.769Z + 0.984XZ \;. \tag{13}$$

Thus, having applied the discriminant procedure, we managed to restore the Rossler system parameters values, Eq. (12), with the accuracy in the range 0.05–0.17.

The accuracy deterioration of the reconstructed factors for the continuous model, as compared with the discrete one, is connected with the numerical differentiation of the initial time series. Filter application allows us to reduce significantly the errors of these fits.

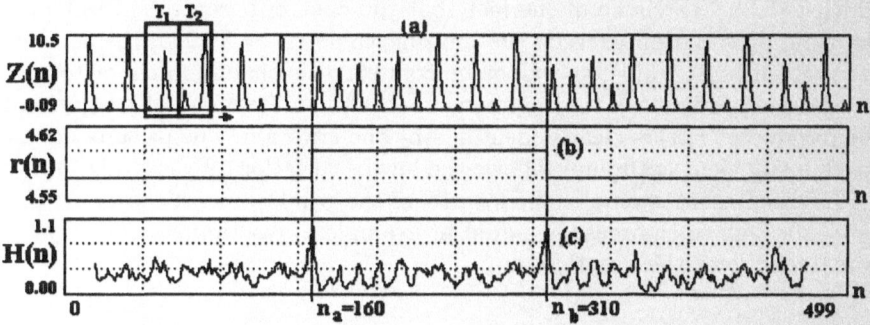

Fig. 2 Discriminant analysis shows sharp peaks of H criterion (c) when the control parameter of the Rossler system changes abruptly (b). Meanwhile the Rossler process itself does not show any parameter variation (a).

If any of the Rossler system parameters changes abruptly at some moment, this can be detected by the H criterion behavior (2) using the discriminant procedure operating with the discriminant function, corresponding to the form of the system equation. Here, H allows one to consistently detect parameter changes less than 0.5%. Let us consider the Rossler system with the control parameter $r(n)$ which changes as in Fig. 2(b). Changes in the Rossler process itself are hardly visible from Fig. 2(a). The last presents only one component, $Z(n)$, from the three component Rossler process. Nevertheless the Fisher criterion H (Fig. 2(c)) demonstrates sufficiently high peaks at the moments n_a and n_b of abrupt changes in parameter r.

2.3 Discrete Systems
with Smoothly Varying Control Parameters

To solve real problems, it becomes necessary to detect smooth parameter changes. The algorithm applicability in such a situation is shown on an example of the discrete system in order to not introduce noise which appears during numerical differentiation.

Figure 3(a) presents sequence $x(n)$ produced by logistic map, Eq. (5), with control parameter, changing according to the law

$$r(n) = 3.81 + 0.1\cos(2\pi n/T) , \tag{14}$$

as shown in Fig. 3(b).

Equation (14) describes the smooth growth of the control parameter $r(n)$ of the logistic map from $r_1 = 3.79$ to $r_2 = 3.81$, during approximately 200 iterations, with a rate which doesn't exceed 0.02 from one iteration (the growth rate is maximal and equal to 0.02 at $n = n_0$). The absolute value of the control parameter change equals t0.02, i.e. about 0.5% of the mean value which is 3.80. By virtue of the fact that the control parameter, throughout the time of variation exceeds the limiting value $r_{cr} = 3.5966$, the sequence $x(n)$, given in Fig. 3(a), as well as in example 1, is chaotic in character.

As seen in Fig. 3(c), the beginning of growth and moment of saturation of the parameter $r(n)$ is clearly identified by the criterion. The parameter reset (with a sufficient accuracy) follows the law of variation (9), (see Fig. 3(d)).

In the present example, the lengths of the windows were chosen as $T_1 = T_2 = 20$. Due to the reasons stated in example 1, discriminant function (6) is matched with the logistic map.

2.4 Discrete Systems
with Periodically Varying Control Parameters

Figure 4(a) presents a sequence $x(n)$ produced by a logistic map with periodically changing control parameter

$$r(n) = 3.81 + 0.1\cos(2\pi n/T) . \tag{15}$$

The law (15) describes the control parameter changes $r(n)$ from the value $r_1 = 3.80$ to $r_2 = 3.82$. Period T was taken as large as 100 (see Fig. 4(b)). Neither Fig. 4(a) nor its Fourier spectrum reveals any visible signatures of periodicity. Due to the fact that here the control parameter $r(n)$ always exceeds its limiting value, the series shown in Fig. 4(a) is always in chaotic conditions. To analyze the series $x(n)$ we made use here, as well as in Examples 1 and 3, of the discriminant function (6) matched with the logistic map.

Meanwhile the discriminant analysis of the above-described type (only with time windows of the same lengths: $T_1 = 8$ and $T_2 = 8$) shows distinct periodicity of H with doubled frequency (Fig. 4(c)). The reset parameter

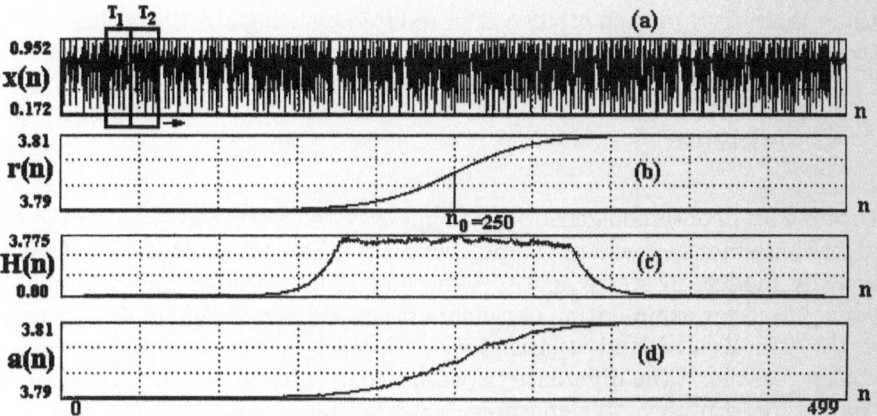

Fig. 3 Logistic map (**a**) corresponding to smoothly varying control parameter (**b**), the action of which is not visible, as well as Fisher criterion $H(n)$ which well identifies the beginning and the end of parameter measurement (**c**), while the coefficient follows its law of variation (**d**).

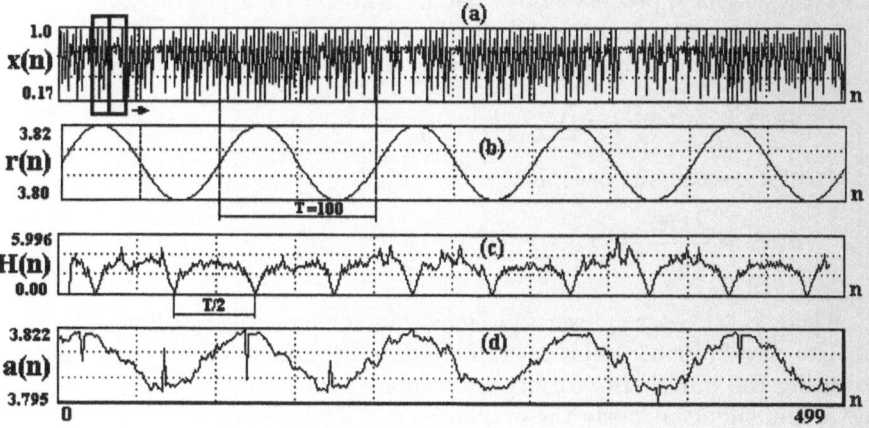

Fig. 4 Logistic map (**a**) corresponding to periodically varying control parameter (**b**). It does not manifest any visible signs of periodicity, whereas discriminant analysis reveals periodicity with doubled frequency in the Fisher criterion $H(n)$ (**c**).

follows the law of variation (15), (see Fig. 4(**d**)). This effect might be of practical interest for communication systems.

The departures (in the form of separate overshoots) of the parameter from the control parameter law for the initial map, occur at moments when the rate of change of $r(n)$ becomes close to zero. Evidently it is connected with poor convergence of the correlation matrices, which occurs when the windows T_1 and T_2 are located near the extrema of the control parameter function. We think that errors of such origin can be reduced by using algorithms insensitive to poor matrix convergence when optimizing the discriminant function.

3 Conclusion

Discrete and continuous chaotic dynamical systems serve as examples of the algorithm efficiency, where the procedure of linear discrimination is used to identify models in states space. A method of discriminant analysis that is widely used for examination of random processes [see, for example, Aivazyan et al., 1989; Kendall and Stuart, 1966] is the statistical method for recognition with a "teacher"; the difference between two or more groups of objects being characterized by several signatures.

Here we used discriminant analysis for solving the inverse problem of non-linear dynamics of nonstationary chaotic systems. The main particularities of adapting the discriminant analysis method from that of random processes to that of nonlinear dynamics are as follows:

1. The dynamical process models determined on two or several adjacent time intervals, are the subjects of the discriminant analysis.
2. The discriminant function is selected for the discrete models as

$$d(\boldsymbol{x}) = \boldsymbol{x}(n+1) - \sum_i \alpha_i F_i[\boldsymbol{x}(n)]$$

and for continuous ones as

$$d(\boldsymbol{x}) = \frac{d\boldsymbol{x}(t)}{dt} - \sum_i \alpha_i F_i[\boldsymbol{x}(t)]$$

where $F_i[\boldsymbol{x}]$ — are power functions (monomials).

The discriminant problem can be reduced to searching for coefficients α_i requiring the minimum of root-mean-square error for d inside each time interval and simultaneously the maximum difference between d values in these intervals. It is exactly this double condition that predetermines, in our opinion, the high sensitivity of the method to the small smooth changes.

The natural further development of this approach is the generalization of vector methods of identification.

Acknowledgment

The authors are very grateful to the International Scientific Foundation for the support of this work within frame of grant NAG300.

References

Aivazyan, S.A., Buchstaber, V.M., Yenyukov, I.S., Meshalkin, L.D. (1989): "Applied Statistics: Classification and reduction of Dimensionality" (Ed. by S.A. Aivazyan). Finansy I statistika, Moscow, 609.

Anosov, O.L., Butkovskii, O.Ya., Isakevich, V.V., Kravtsov, Yu.A. (1995): "Nonstationarities revealed in random-like signals of dynamical nature". Radiotechnika i Electronika **40**(2), 255–260. [transl. Sov. J. Communication Technology Electronics, 1995, **40**(2)].

Breeden, J., and Hübler, A. (1990): "Reconstructing equations of motion from experimental data with unobserved variables". Phys. Rev. A **42**(10), 5817–5826.

Brush, J.S., and Kadtke, J.B. (1992): "Nonlinear signal processing using empirical global dynamical equations". *Proceedings of the ICASSP-92*. San-Francisco, 321–325.

Gouesbet, G. (1991): "Reconstruction of the vector fields of continuous dynamical systems from numerical scalar time series". Phys. Rev. A **43**(10), 5321–5331.

Gribkov, D.A., Gribkova, V.V., Kravtsov, Yu.A., Kuznetsov, Yu.I., Rzhanov, A.G. (1994a): "Reconstruction of dynamical system structure from time series". Radiotekhnika I Elektronika **39**(2), 241–250 [transl. in: Sov. J. Commun. Technol. Electron., 1994, **39**(2)].

Kendall, M.G., Stuart, A. (1966): "The Advanced Theory of Statistics". v.3. Charles Griffin and Company Limited, London, 736.

Kravtsov, Yu.A. (1989): "Randomness, determinateness, predictability". Sov. Phys. Uspekhi, 1989, **32**(5), 434–449.

Kravtsov, Yu.A., Ed. (1993): "Limits of predictability". Springer Verlag, Berlin, Heidelberg.

Ljung, L. (1987): "System identification: theory for the user". Prentice-Hall, Englewood Cliffs, N.J.

Mason, J., Treas, R.S. (1986): "Numerical weather prediction". *Proc. Roy. Soc.* A**407**(1832), 51–63.

Classifying Complex, Deterministic Signals

James B. Kadtke and Michael Kremliovsky

Abstract

When attempting to analyze, model, or predict time series, one often finds
that the data is so complex or noisy that the presence or location of a desired
signal or signals is unknown. Therefore, the first step in such an analysis is often to implement some scheme for detecting or classifying signals in otherwise
long stretches of noisy background. Detection/classification of signals is thus
one of the principal areas of signal processing, and the utilization of nonlinear
information has long been considered as a way of improving performance beyond standard linear (e.g., spectral) techniques. Here, we develop a method
for using global models of chaotic dynamical systems theory to define a signal
classification processing chain which is sensitive to nonlinear correlations in
the data. We use it to demonstrate classification in high noise regimes, using
short data segments which mimic real-world processing restrictions, and also
show that classification probabilities can be directly computed from ensemble statistics in the model coefficient space. We also develop a modification
for non-stationary signals (i.e. pulses) using non-autonomous ODEs. Finally,
we demonstrate the technique by analyzing actual open ocean acoustic data
from marine biological sources such as whales and dolphins.

1 Introduction

Prediction and modeling of complicated time series is an important problem.
However, in many cases the data are so complex or noise corrupted that one
may wonder whether there is a deterministic component present at all. In such
cases, the first step in an analysis may be to determine if a "signal" exists at
all, and if so where, and secondly, whether any identified signals are generated
by the same system, or whether they are realizations from several separate
signal classes. In this contribution, we will thus be concerned with developing
some new numerical tools for the detection/classification of signals, or, to
be more exact, signal components which we shall assume are generated by
some deterministic process. Detection and classification form one of the most
important areas of the field of signal processing, and indeed come under
the guise of many names in many fields which deal with the problem of
determining structure in data. Roughly, signal detection means determining
the existence and location of a (usually) deterministic signal-of-interest (SOI)
in an otherwise (possibly) structureless data set, while classification generally

assumes that the presence and location of the signal are known, but desires to assign the detected signal to one of several signal classes usually defined beforehand. Note that these problems are typically posed for signals which are highly noisy, since detection is not really an issue in an otherwise noise-free data set. Classification can be an issue in the low-noise regime, however, if the particular signal classes, e.g. non-stationary and nonlinear, are such that they are not well described (and hence poorly discriminated) using typical signal measures such as power spectra.

The interest in algorithmic (i.e., automated) detection and classification techniques has grown enormously in the last decade, due to the increased use of sensor technologies in everyday life and the inability of human operators to process the growing volume of data. The applications of detection and classification methods go well beyond the obvious ones, however, such as sonar and radar signal analysis. Voice and pattern recognition, and related security problems, comprise a large area of interest. Biomedicine may provide one of the most important new applications, for example in looking for new ways of characterizing physiological states using data from EEG, EKG, etc. Analysis of geophysical data may also provide a similar application, for example in attempting to isolate precursors to earthquakes. The analysis of higher-dimensional data sets, e.g. spatio-temporal data, is also more plausible now, due to enormous increases in computational speed.

Here, we will develop a method which attempts to classify signals using ideas derived from the theory of Nonlinear Dynamics, or chaos. That is, we will assume that the deterministic component of the SOI can be described by some dynamical system, which is possibly nonlinear and multivariate, and that the data we measure can be assumed to be some scalar physical observable of this system. The use of a "state space" to describe data is not new; however, nonlinear dynamics assumes a particular model class for this application that is somewhat novel: that the time evolution of the data can be described as a dynamical system, or rather a system of differential equations describing the behavior of dynamical modes in the physical system (a "global dynamical model"). Note that this approach was put on rigorous grounds by Takens theorem [Takens, 1980], and hence is comparatively recent. Takens theorem allows us to measure a single scalar observable from the system, and attempt to recover or model its dynamical properties in a reconstructed time-delayed state space, because certain topologically invariant quantities are preserved. Hence, if we choose to build a model of a system (or data) which we assume was originally derived from a dynamical system, then in a sense an empirical dynamical system fit to the data is the "correct" model class, and we have a chance of optimally capturing the data properties.

Note that one can take two philosophical approaches to a data analysis problem using these methods: first, one could ignore entirely the issue of whether there actually *is* a dynamical system generating the signal, and rather only assume the data can in fact be modeled as one. In this case, one takes the "black box" approach, and the description is purely empirical (which in many practical applications is entirely sufficient). However, if we deal with physical systems which we suspect may actually evolve according to some dynamical system, we may hope that a dynamical model may capture something more fundamental about the physical process, and hence the structure of the estimated model may have deeper meaning.

From a signal processing point of view, the time series methods of nonlinear dynamics are innovative because they allow a compact representation of nonlinear signal information (i.e., nonlinear correlations) to all orders, since this information is captured in the couplings between dynamical modes. Although powerful, conventional linear methods, e.g., matched filtering, are largely insensitive to nonlinear correlations or can distort them [Badii and Politi, 1986], and hence do not make use of the additional nonlinear information which may be present in many natural data sets. The use of nonlinear correlations for detection/classification is not new (e.g., higher-order spectral methods, first proposed in the 1960s; see also [Elgar and Chandran, 1993]); however, these methods can have exorbitant data requirements or perform poorly in the high noise regime, and most importantly capture only one moment of the nonlinear expansion at a time. Nonlinear correlations can, however, be efficiently represented to all orders in the reconstructed dynamical phase space.

Time series analysis methods derived from nonlinear dynamics have been applied to a variety of signal processing tasks in the last decade [Farmer and Sidorovich, 1993; Casdagli, 1989; Hammel, 1990; Kostelich and Yorke, 1990; Cawley and Hsu, 1991; Pecora and Carroll, 1990; Hayes et al., 1994]. In recent years, interest has developed in practical uses for dynamical models as detectors of determinism, and even as classifiers [for example, see Mindlin and Gilmore, 1992]. Unfortunately, many of these methods require a priori exact signal models or clean (i.e. noise-free) signal realizations, or perform well only at modest noise levels. Also, few methods incorporate the dynamical information into a useful classification chain, yielding classification probabilities as an output [an exception is recent work by Fraser, 1995]. Here, however, we develop a very general detection/classification method which has several advantages from a practical, operational point of view [Kadtke, 1995]. Even though we model signals as a dynamical system, this method does not assume signal stationarity or an "attractor" over the measurement time scales, and so is equally applicable to linear (e.g. tonals) and nonlinear signal components, and stationary or transient signals. Because of the averaging characteristics of the model class (sets of ODEs), this technique can perform well in the high-noise regime. In this paper, we define the high noise regime to be the negative decibel (dB) range, with SNR $= 20\log_{10}(\sigma_{signal}/\sigma_{noise})$, σ standard deviation, since this is the range where most conventional detection/classification schemes have difficulty. This technique also does not assume the existence of any exact model for the SOI, nor does it initially require clean realizations of particular signal classes to be useful. It also provides a simple way of calculating the classification probabilities, and hence performance can be estimated in a straightforward fashion. Finally, the method is numerically efficient and the data requirements are minimal, and hence the scheme is likely applicable to real-time data processing.

The inclusion of a nonlinear dynamics technique into a conventional detection/classification processing scheme can often be confusing, and it is useful to consider the scheme as a sequence of distinct steps. In the following technical discussion, for conceptual organization, we divide the detection/classification process as follows: we choose first the type of data representation, e.g. scalar data, or time-delay reconstructed vector data; second, an information quantifier, such as a power spectrum, or the fitting of a dynamical model; third, an information metric, which is capable of assigning values to differences in quantifier realizations in a useful fashion; and finally a detection/classification statistic, which is typically a function of the metric. In this view, it is possible to interchange different numerical approaches for each step, and hence form hybrid schemes. We have investigated several such hybrid schemes, and the scheme we present below was generally found to be optimal for most simulated data tests.

In the remainder of the paper, we will first discuss the theoretical and numerical aspects of the detection/classification scheme, giving some numerical examples on simulated data. We then discuss some of the technical issues concerning its application to non-ideal (real-world) data sets. The Sect. 4 then presents an example of its application on a real-world data set. Finally, we draw conclusions and indicate future directions for the research.

2 Dynamical Classification Algorithm

2.1 The Algorithm

We develop our detection/classification scheme by first discussing the algorithmic procedure, then discuss how it is operationally applied to a given data set. To begin, we assume the existence of a time series $x(t)$ of measurements of some generic physical observable of a system, or of an artificially generated signal. We use as our data representation the time-delay reconstructed state-space vectors $x = (\ x(t),\ x(t-\tau), \ldots,\ x(t-(D-1)\tau)$ as per Takens [Takens, 1981], for some embedding dimension D and delay τ. Intuitively, this step constructs from the scalar data $x(t)$ a higher-dimensional dynamical state-space representation with potentially D degrees of freedom, isomorphic to the original phase space for sufficient dimension. As is well known, a deterministic signal evolution (even for linear systems) in this reconstructed state-space can be described by a system of D first-order differential equations, which may be coupled and nonlinear. The signal evolution generates state-space trajectories whose time-asymptotic distribution can be characterized by different properties: global topology (the fixed-point structure and manifold geometry); determinism (existence of a dynamical "rule"); differentiability (for continuous systems); and distributional (e.g., invariant probability density). We attempt to develop a classification scheme which is sensitive to all these various types of state-space characteristics simultaneously, even though much of it is often obscured at high noise levels.

Therefore, we choose as our information quantifier an empirically fit global dynamical model [Crutchfield and McNamara, 1987; Cremers and Huebler, 1987]. That is, we fit a set of model differential equations, $d\boldsymbol{x}(t)/dt = \boldsymbol{F}(\boldsymbol{x}, t)$, to the data evolution in the reconstructed state space vectors $\boldsymbol{x}$. As a fundamental point, we assert that if the original data generator was actually a continuous dynamical system, then a dynamical model is of the "correct" model class, and can be simultaneously sensitive to all the types of dynamical information mentioned above. That is, in the limit of vector observables, infinitely long time series, no noise, and the proper basis expansion, we can recover the original dynamical system to arbitrary accuracy. In practice, for detection/classification we are not concerned with the exact model structure, and we construct a model by assuming some general expansion for the D functions $\boldsymbol{F}(\boldsymbol{x}, t)$ in some basis set $\phi(\boldsymbol{x}(t))$. Typically, for the autonomous case $\boldsymbol{F}(\boldsymbol{x}, t) = \boldsymbol{F}(\boldsymbol{x})$, one can simply choose all polynomials up to some order P, with $\phi_j(\boldsymbol{x}(t)) = x(t)^k \cdot x^m(t - \tau) \cdots \cdots x^q(t - (D-1)\tau)$, where $0 \leq k$, m, $q \ldots \leq P$ and $k + m + q + \ldots \leq P$, writing for each of the $1 \leq d \leq D$ equations

$$\boldsymbol{F}_d(\boldsymbol{x}, t) = a_{0,d} + a_{1,d}x(t) + \ldots = \sum a_{j,d}\phi(\boldsymbol{x}(t)) , \quad j = 0, \ldots N \qquad (1)$$

and the expansion includes all $N = (P + D)!/P!D!$ cross terms. The unknown model coefficients $a_{j,d}$ are determined by constructing a general linear least-squares problem [for example, see Press et al., 1986] for each F_d defined by the matrix equation $\boldsymbol{Y} = \boldsymbol{B}\boldsymbol{A}$, where $\boldsymbol{Y}$ is the vector of time derivatives $dx(t)/dt$ estimated at successive observation times $x(1)$, $x(2), \ldots$; B is a matrix of the basis terms $\phi_j(\boldsymbol{x}(t))$ evaluated at $x(1)$, $x(2), \ldots$ (i.e., the "data matrix"); and A is the vector of unknown coefficients $a_{j,d}$. Each F_d is determined by least-squares solution of its matrix equation using, e.g., singular value decomposition. Many papers have appeared in the literature in recent years discussing various numerical approaches to this dynamical *inverse* problem, and the reader is referred there. For prediction and modeling purposes, nontrivial numerical issues are still outstanding [Kadtke and Brush, 1993]; however, it has been shown that recovery of a system's dynamical invariants can often be achieved even for significant noise [Kadtke, Brush and Holzfuss, 1993; Brown, 1993].

It is important to note once again that, for detection/classification purposes, our scheme diverges fundamentally from a modeling approach. For modeling purposes determination of a correct D, P, etc. is necessary to recover exact dynamical information about the original system, which is typically problematic. However for detection/classification purposes, we postulate that it is only necessary to incorporate sufficient D and P (and other parameter ranges) to distinguish the required signal classes to a particular performance level, regardless of the exact form of the original dynamical generators. Typically, we find that D and P can usually be small (e.g., 2 or 3), since signal power is usually distributed mostly in the lowest orders of the model coefficients (in analogy with spectral expansions). More importantly, we find it crucial to define a standard form (i.e., fixed P, D, τ) for the dynamical model (i.e., a fixed "dynamical filter") in a particular application, otherwise comparison of arbitrary signal classes which may have radically different forms for their generators is impossible.

We now assume that we have extracted estimated coefficient sets for a variety of signal observations α. There now remains the non-trivial issue of quantifying differences in coefficient distributions in a meaningful way (i.e. the *information metric*). We use a simple but useful scheme: we construct a vector space $\mathcal{A}$ from the coefficients of all $\phi_j(\boldsymbol{x}(t))$ by forming vectors $\boldsymbol{A}_\alpha = (a_{0,1},\ a_{1,1}, \ldots,\ a_{j,d}, \ldots)$ for all α data realizations, where the $a_{j,d}$ include all $D \cdot N$ available coefficients from the D model functions F_d. For each $\boldsymbol{A}_\alpha$, the $a_{j,d}$ determined from the fitting process are embedded in this high-dimensional coefficient space $\mathcal{A}$. Note that each $\boldsymbol{A}_\alpha$ in $\mathcal{A}$ defines a unique model dynamical system, describing a particular data realization, and the set of all $\boldsymbol{A}_\alpha$ for an ensemble of observations yields a corresponding distribution in $\mathcal{A}$. Qualitative differences between the different signal realizations can then be quantified using an Euclidean metric to measure differences in the $\boldsymbol{A}_\alpha$.

A central part of any detection/classification scheme is the calculation of probabilities and confidence estimates. Using the representation $\mathcal{A}$, classification probability estimates can be directly derived from the statistical properties of the ensemble distributions in $\mathcal{A}$. That is, the probability that the αth data observation belongs to the mth signal class can be estimated from the probability that a particular $\boldsymbol{A}$ lies within the ensemble distribution C_m of that class in $\mathcal{A}$. For C_m which are Gaussian, this probability is directly estimated from the distance of $\boldsymbol{A}_\alpha$ to the centroid of C_m, normalized in units of the standard deviation of C_m. In practice, the C_m are typically non-Gaussian, hence we utilize non-parametric statistics to estimate probabilities. In this context, we note that pure noise must generate Gaussian C_m centered at the origin of $\mathcal{A}$, and that here the "detection" process consists simply of the statistical exclusion of a given $\boldsymbol{A}_\alpha$ from the C_m of the noise realization. Hence, we may perform detection/classification simultaneously, if we are given prior (noisy or clean) realizations of the SOI and the pure noise background (which may be colored).

The above discussion describes the algorithmic tools used in the scheme, and we now describe how they are used operationally. Typically, we will be interested in either transient signals which have a short, finite lifetime, or short observations of a potentially stationary signal, both of which can effectively limit the window-of-observation to a few characteristic timescales. This, coupled with the fact that the model basis will generically be incorrect, means that even noise-free signal observations will produce a statistical distribution in the coefficient estimates for the A_α. Hence, one must resign oneself to the idea that even repeated observation of an identical signal will generically produce a distribution in $\mathcal{A}$, and only a probabilistic estimate for classification can be made. These facts imply two things about the technique: first, the details of the observation procedure become important in the coefficient estimation scheme, i.e. the sampling rate, the data observation window length L_w (in samples), and number of observations, among others, must be accounted for and standardized in the definition of the "dynamical filter" used in the detection/classification process. Secondly, characterization of the statistical properties of the coefficient ensembles in $\mathcal{A}$ is a natural and useful way to estimate the probabilities of the finite-observation-time effects, and provides a natural metric for studying the differences of signal classes. Therefore, we will think operationally in terms of collecting statistics on model estimates of many short windows of independent observations of signals and noise, then using these ensembles to perform detection/classification on future observations of arbitrary signals.

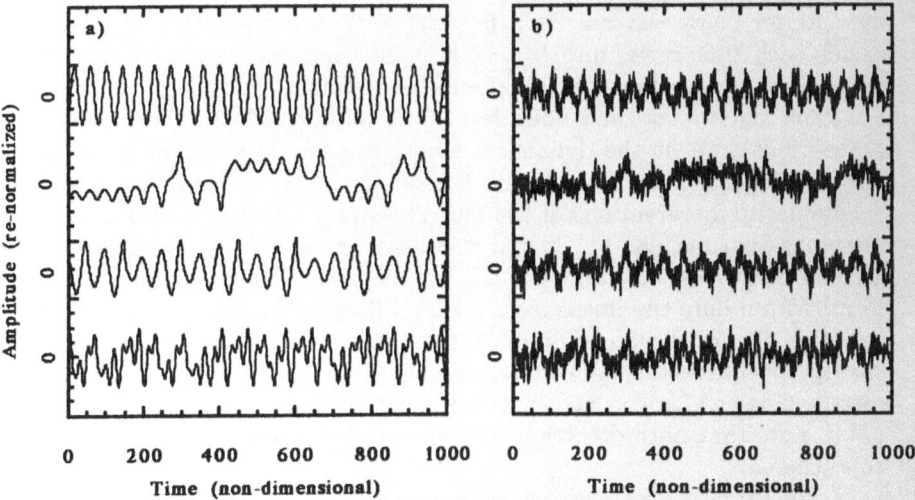

Fig. 1 Segments of pure (*left*) and 0 dB noisy (*right*) time series, from sine wave (*top*), Lorenz system, Rossler system, and Pecora circuit (*bottom*).

2.2 Numerical Tests

We now clarify the above ideas using several numerical examples. Throughout, we will explicitly construct our simulated data to minimize or remove amplitude and spectral information, hence detection/classification results are due principally to nonlinear correlation. We first simulate data observations from a stationary system by generating long time series $x(t)$ from three systems: a sine wave, and the Rossler and Lorenz chaotic dynamical systems; we then also include real data from the Pecora chaotic circuit [Pecora and Carroll, 1990]. All series simulate discrete sampling at 50 to 100 points per characteristic "orbital" time, which normalizes the time scales so that power spectral distributions for the different systems overlap. Simulated Gaussian-distributed noise was then added at different noise levels from $+5$ to -10 dB. All time series were then normalized to zero-mean and unit variance throughout (via a windowed average) to remove amplitude information and make conventional signal variance detectors insensitive. Segments of resulting pure and noisy signals are shown in Fig. 1, left and 1, right, respectively. Power spectra for the broadband signals (not shown) yield little structure for negative SNR, hence matched filter techniques are also not useful.

To simulate a detection/classification procedure, we choose a standard dynamical model with $D = 3$, $P = 2$, and $\tau = 12$ ($\approx 1/4$ cycle). This system has 30 total coefficients $a_{j,d}$ in $D = 3$ time-delayed variables, which we re-write for clarity as $x = x(t)$, $y = x(t - \tau)$, and $z = x(t - 2\tau)$. We then divide each time series into $L_w = 500$ point windows (about 10 characteristic cycles), to simulate individual and independent short data observations. For each signal class and noise level (and additionally for pure noise data realizations), we fit the dynamical model to each data window, generating the A_α. In this sense, we "train" the detection/classification algorithm by accumulating observations of the four known signal classes, and a "library" of signal distributions C_m is stored consisting of the A_α. For the data described above, Fig. 2 shows the C_m for the 0 dB SNR cases, in the 3-space formed from only the linear basis term coefficients $a_{1,1}$, $a_{2,1}$, and $a_{3,1}$ of one equation F_1, for visualization purposes. Clear separation is evident between the C_m for different signal classes and noise, except for the Lorenz and Pecora distributions which can be separated using alternate axes. As an important point, note that both detection/classification are achieved simultaneously in this scheme.

As mentioned previously, detection/classification probabilities can be assigned using the statistical properties of the C_m, such as those shown in Fig. 2. We have found that the C_m for deterministic signals are usually significantly non-Gaussian, and tightly clustered about the centroid. Often more than 90% of the A_α lie within one equivalent σ of the mean, hence higher confidence estimates can be achieved than would be indicated by Gaussian statistics. To illustrate this, Fig. 3(a) shows a sample distribution of a C_m along only one coefficient axis, for the Rossler system at five representative

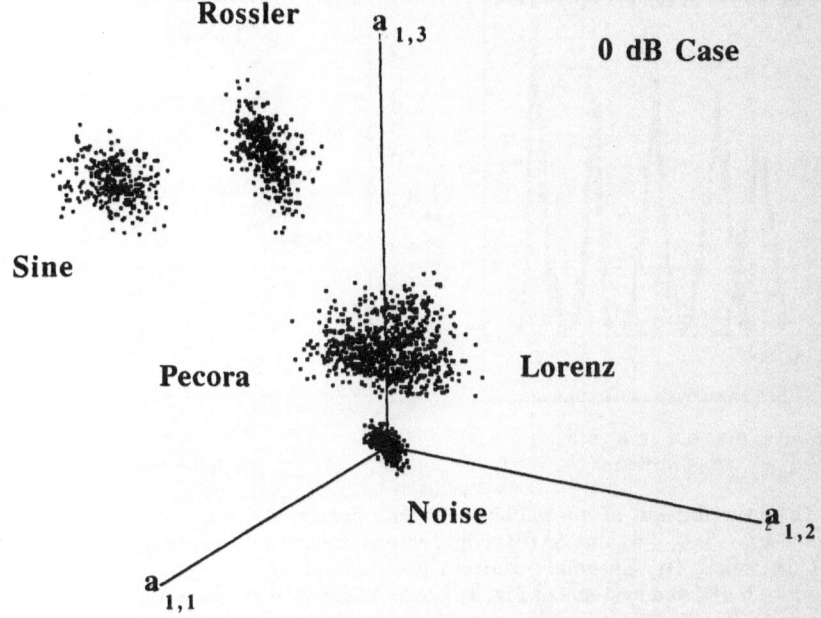

Fig. 2 Three-variable coefficient space ($a_{1,1}$, $a_{2,1}$, $a_{3,1}$) showing embedded ensembles of model coefficient realizations, for signals of Fig. 1.

SNR levels. As is evident, the distribution is tightly clustered at larger SNR, and becomes more Gaussian for lower SNR, while the centroid simultaneously moves toward the origin of $\mathcal{A}$.

The above analysis has assumed data at a constant, known SNR level. In practice, SNR levels are often unknown and individual data observations may appear at arbitrary SNR. Importantly, we have demonstrated that systematic scaling exists for the ensemble centroid positions with varying SNR level. This is demonstrated by numerical experiments in Fig. 3(b), showing the scaling with SNR of the centroids of several signal classes, projected along two axes of $\mathcal{A}$ for visualizaton. As the SNR varies, the centroid positions smoothly trace a line which moves to the origin of $\mathcal{A}$ as the SNR moves to $-\infty$ (note that these lines are significantly more separated when additional axes of $\mathcal{A}$ are used; also, all lie in the same quadrant due to artificial time normalization). These results indicate that it is possible to perform classification for arbitrary noise level, by having prior signal observations at only a few SNR, and then extrapolating the SNR-scaling of the centroid. Classification statistics are then measured from this SNR-parameterized line, rather than a single position at a specific noise level. Note that Fig. 3(b) indicates that detection/classification is possible in these cases even down to -10 dB.

We now make the point that in practice, reliable detection/classification can usually be achieved using relatively few axes of $\mathcal{A}$. Firstly, we note that for

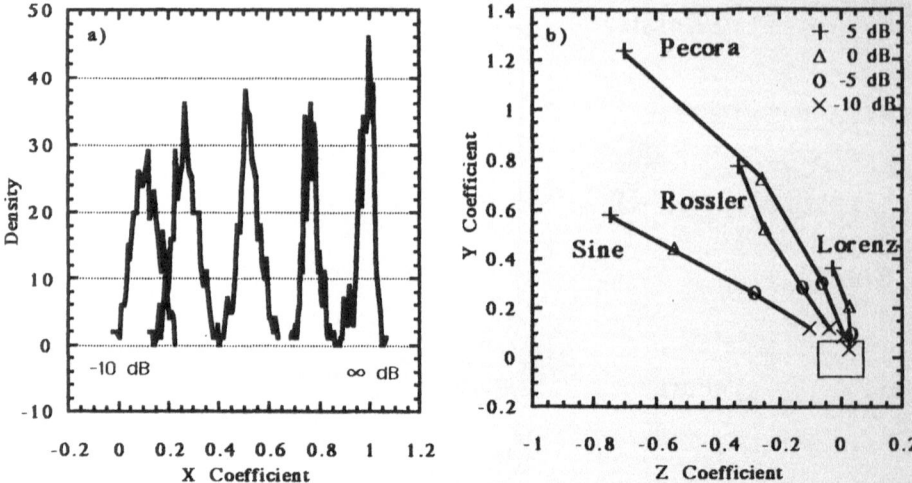

Fig. 3 (a) Distribution of ensemble coefficient densities along one axis of $\mathcal{A}$ for SNRs of -10, -5, 0, $+5$, and ∞ dB. Highly peaked distributions become Gaussian as SNR decreases. (b) Ensemble centroid positions along two axes of $\mathcal{A}$, for -10, -5, 0, and $+5$ dB, and systems of Fig. 1. Box is origin of $\mathcal{A} \pm$ one noise σ. Centroids trace smooth line to origin as SNR decreases.

time-delay reconstructed data, the Toeplitz nature of the data matrix implies that all equations F_d contain qualitatively similar or redundant information, particularly in the high-noise regime, and hence we need only monitor the $a_{j,d}$ of one of them. Secondly, for any given signal class, we typically find that few coefficients contain statistically significant information, the rest reflecting only noise properties. We find that these significant coefficients can be identified as follows: we renormalize all coefficients $a_{j,d}$ by their standard deviations $\sigma_{j,d}$ estimated from the C_m of the given class. We then argue that the new variables $a_{j,d}/\sigma_{j,d}$ that lie within one or two $\sigma_{j,d}$ of zero contain little or no signal information, since they statistically include zero. We thus define the significant coefficients as those lying outside this threshhold, and these must scale with noise magnitude. As an example, Fig. 4 shows a plot of the lowest order $a_{j,d}/\sigma_{j,d}$ vs. coefficient number for the Rossler signals of Fig. 1, for varying SNR value. The one-σ threshold is also indicated. Here, the two statistically significant coefficients characteristic of this signal class are easily identified. This property is generally valuable since restriction of the detection/classification statistic to the significant coefficients can decrease the probability of false identification.

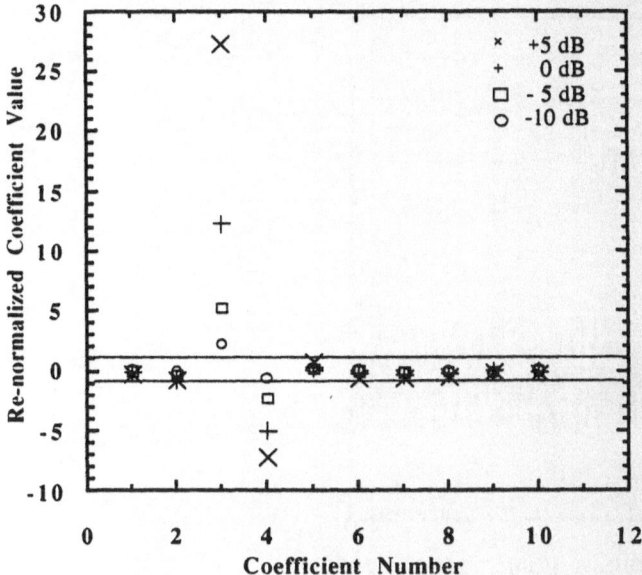

Fig. 4 Normalized coefficient values $a_{j,d}/\sigma_{j,d}$ for Rossler data, at different SNR. Only two coefficients lie outside one-sigma threshold (indicated by lines) and are statistically significant.

2.3 A Non-autonomous Formulation

The numerical experiments above demonstrate the general detection/classification procedure, by using short windows of otherwise stationary signals. A more difficult problem is encountered with non-stationary signals, where the useful observation length is limited by the characteristic signal time. In particular, detection/classification of short-lived, pulse-like signals (i.e. "transients") is important in many applications, and few reliable methods exist, especially for high noise regimes. We now demonstrate detection/classification of transients in high noise using a simple modification of the previous technique. We modify our technique solely by explicitly including time-dependent functions in the global model, i.e. we make the $F(x)$ of (1) non-autonomous by including time-local basis terms. Coefficient estimation and representation in $\mathcal{A}$ are then performed exactly as before. We illustrate this by the following numerical experiment: we generate a simulated transient signal from Rossler data using exponential enveloping, i.e. $x(t) = \exp(-|t-t_0|/\lambda)R(t)$, with $R(t)$ a Rossler time series segment from Fig. 1. We choose λ so that the pulse is "short", i.e. it contains < 10 characteristic oscillations of the Rossler dynamics. We then add Gaussian-distributed simulated noise at differing SNR between $+5$ and -10 dB. Figure 5 shows samples of pure and noisy (0 dB) single pulse realizations. We now define the $F(x,t)$ to be

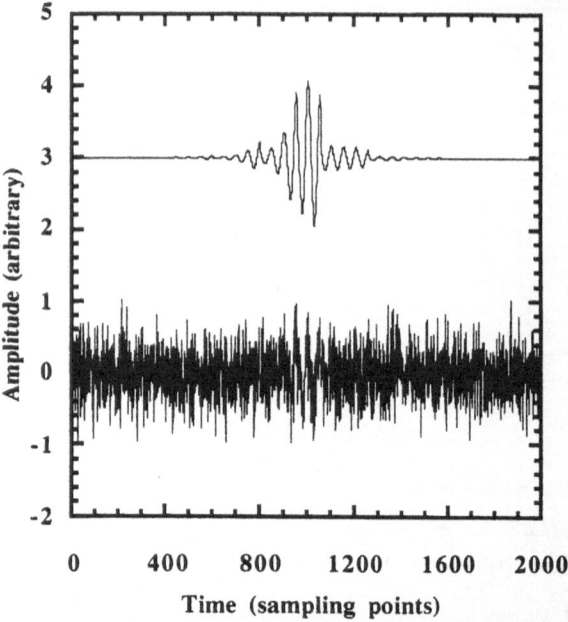

Fig. 5 Time series of simulated Rossler chaotic "pulse" at ∞ dB (*top*) and 0 dB SNR (*bottom*).

$$dx(t)/dt = a_{0,1}x(t)(t - t_0) + a_{1,1}x(t) + a_{2,1}y(t) + \ldots$$
$$dy(t)/dt = a_{0,2}y(t)(t - t_0) + a_{1,2}x(t) + a_{2,2}y(t) + \ldots \tag{2}$$

i.e., a model with $D = 2$, $P = 2$ and one additional time-dependent term per equation. Note that eq. 2 contains as a possible solution

$$x(t) = \cos(bt)\exp(-|t - t_0|^2/\lambda) , \tag{3}$$

i.e. a pulse-like solution, but of a different form than the Rossler envelope. We then fit (2) in a sliding $L_w = 150$ point window over the signal, at the various noise levels, and monitor the coefficients in $\mathcal{A}$ as before. A plot of typical results is shown in Fig. 6. Here, we plot for visualization purposes the value of only one coefficient $a_{1,2}$ (corresponding to a re-scaled frequency) vrs. time (window number). Detection of the pulse is evident even for SNR of -5 dB, and statistical detection occurs down to nearly -10 dB; slightly better performance is obtained using additional axes of $\mathcal{A}$. As a check, we phase randomize [Theiler, 1992] the pulse data to destroy nonlinear correlations, and repeat the experiment. The resulting plot (bottom curve, Fig. 6) shows no variation over the pulse, hence we infer that the detection/classification ability is due principally to nonlinear correlations.

To summarize this experiment, we have demonstrated detection of a short-lived, highly non-stationary, strongly nonlinear signal in a high noise regime,

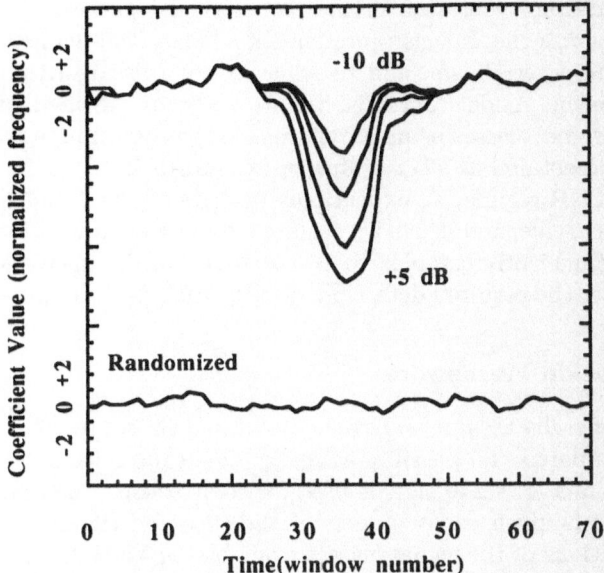

Fig. 6 Detection of pulse of Fig. 5 using windowed non-autonomous model. Plot shows one coefficient (re-scaled frequency) vs. time (i.e. window number), for four SNR levels. Bottom curve is same experiment for phase-randomized data.

even while using the "incorrect" dynamical model to perform coefficient estimation. By comparison, most conventional schemes show poor performance for these signals for 0 dB and lower.

3 Some Technical Issues for Real-World Data

In the last section of this paper, we described a nonlinear detection/classification scheme using global dynamical models which was able to detect and classify non-stationary, noisy (−10 dB), and typically short data observations in a numerically efficient way, which however was demonstrated only on simulated data sets. In the next two sections, we turn to the analysis of real-world data sets, to examine its practical usefulness. Obviously, the analysis of real data presents new problems, often yielding ambiguous results, and we discuss these issues in some detail. In general, when performing any such analysis, one must make several assumptions about the character of the SOI and its representation as a dynamical system, and also about the parameter ranges which may be relevant to the particular time scales of the problem. We will briefly cover, in an outline form, these issues below. One general point is that, although we attempt to perform detection/classification using a low-dimensional nonlinear dynamical model, we generally do not require this property of the particular signal class we analyze. We typically require only stability of the signal representation in the reconstructed phase space (i.e. non-divergent), otherwise the Takens representation is not well-defined. We of course also assume determinism, and to some extent this algorithm is a detector of determinism. Aside from this, we are certainly justified in representing a variety of signal classes using a dynamical representation, e.g., a tonal has a simple representaion as a set of linear, two-dimensional ODEs. In particular, in the low SNR regime, we expect only vestiges of any detailed phase space structure to survive, and hence we require a dynamical model for the classifier to be only sufficiently complex to be sensitive to this surviving information, and to deliver the required detection/classification performance.

3.1 General Classification Framework

Classification schemes generally assume a known set of signal classes to which an arbitrary signal is compared and possibly assigned. One valid question is how one obtains the "canonical" set of signals used for comparison. One may consider cases where one is given exact models of the expected signals, or clean (zero noise) realizations of the signal classes available, or know the exact statistical properties of the background noise (white, colored), etc. Here, we assume no a priori fundamental knowledge of the signal types, or a detailed understanding of the characteristic noise spectrum. Hence, we attempt to perform only relative detection/classification, in the sense that we will observe an ensemble of realizations of noise and noise + SOI, hence building a "library" of reference signals. Note that, in a sense, this consists of "training" our classifier. We then attempt to classify new, arbitrary signal observations based on the structure (partitioning) in coefficient space of these known signal classes. In this paper, we will not present a detailed classification space analysis nor derive statistical estimates, which can be complicated issues, but rather we will simply demonstrate separation of the three signal classes in the model coefficient space.

3.2 Choice of a "Dynamical Filter"

As was discussed in Sect. 2, we define a fixed "dynamical filter" to extract signal information, through which all signal classes are viewed. This filter is, in our case, a set of nonlinear ODEs defined in the re-constructed Takens state space, and hence the system is defined by a dimension D and an order P. Larger D can have an effect on the noise averaging characteristics (and "over-embedding" methods can even be implemented), and in general we need at least $D = 2$ to represent a tonal, so a typical choice is $1 < D < 6$. However, for low SNR and a limited data window length, higher D can result in very sparse data and correspondingly poor numerical results. For the purposes of this data set, we chose $D = 3$ based on noise level. Similarly, a larger P can capture much more signal complexity and hence improve classification ability. However in the presence of strong noise this information is typically minimal, and higher P can actually increase the model instability, again because of data sparseness. Generally, we limit ourselves to $1 < P < 5$, and here, because of the low SNR and short window lengths, we chose $P = 2$.

3.3 Choice of Delay Time

The delay time τ used to construct the Takens vectors is an important parameter, although the results are typically stable within a significant range once this is determined. For clean data from well-defined dynamical generators, one can obtain a good estimate for τ using, e.g. mutual information. For very noisy data, and in particular for non-stationary data, the issue is significantly more complicated. One guideline which we have found useful is to simply use approximately 1/4 of the period of the characteristic time scale of the pulse oscillation, which generally gives good results. Since most of our data here lies between 40 to 50 points per characteristic period, we choose a $\tau = 12$. Generally, one must also ensure that the chosen τ is longer than any correlation time which may be artificially induced by the analysis procedure, for example, the smoothing procedure used to determine the local derivative.

3.4 Choice of Data Windows

For a D-dimensional system, one requires D ordinary differential equations to completely describe the time evolution. However one can argue that, because of symmetries induced by the Takens construction and also the loss of detailed information in the high noise regime, that in actuality each separate ODE contains qualitatively similar information from a detection/classification performance standpoint. Hence, we typically utilize only one equation, which comprises our dynamical filter, of the form

$$\frac{d}{dt}x(t + (d_e - 1)\tau) = c_0 + c_1 x(t) + c_2 x(t + \tau) + c_3 x(t + 2\tau) + \ldots +$$
$$+ c_{d_e} x(t + (d_e - 1)\tau) + \text{high order terms} \tag{3}$$

This filter effectively maps a single window of the data into the coefficient space constructed from the $\{c\}_i$. Many observations of a given signal class are fit in this way, and the resulting coefficients from each window form an ensemble in the coefficient space. The variations in this ensemble are due to the noise component, the short obervation time, least-squares numerical variation, etc. For stationary signals, we can simulate many individual observations by sliding the data window along a single time series. Thus we have one and potentially two additional parameters to specify: the length of the window, L_w, and the shift of the window, L_s (in the latter case). Some guidelines can be made concerning choice of these two parameters: if L_w is too long in the case of a low SNR signal, then it is difficult for the filter (1) to "adjust" itself self-consistently over the whole data window. In other words, there are too many singular equations after singular value decomposition (SVD) of the matrix equation. In the case of a short window ($L_w \sim T$), there is not enough information to specify the topological pecularities. Typically, we find that the optimum value can be found in the range $2T \leq L_w \leq 6T$, where T is the characteristic time of the signal. In the case of sliding windows for a stationary data set, we generally put $L_s = L_w/2$, which makes the windows statistically independent, but usually allows us to increase the density of the ensemble in coefficient space.

3.5 Estimation of Local Derivatives

One of the most important elements of fitting the model dynamical equations is the estimation of the left-hand side of the matrix equation, consisting of the local first-derivatives at some data point. Success of the detection/classification scheme strongly depends on whether we are able to extract continuous time derivatives from a badly corrupted (i.e. noisy) signal. To some extent, we assume and utilize the property that determinism is equal to differentiability [Salvino and Cawley, 1994]. It must be stressed however that for detection/classification, continuity of the derivative is more important than its true value, because the latter can be renormalized or even smoothly transformed without loss of generality. What is important is that random data produce random derivatives (on average), which in essence defines a null hypothesis for the technique. In practice, one must always define some smoothing procedure for noise corrupted (i.e. non-differentiable) data, and we utilize a least-squares interpolation procedure of quadratic order which averages over L_{sm} number of data points before and after the point in question. Since this smoothing procedure itself may induce artificial structuring of the data, one must avoid these problems by judicious choice of time scales. This is usually most easily accomplished by requiring that τ is at least as large as L_{sm}. Here, for this rather highly noisy data, we will use a L_{sm} of 8, and always choose τ at least this large.

3.6 Simple Statistical Estimate of Classification Probabilities

Any detection/classification procedure must be concerned with estimation of the probabilities and confidences associated with any class assignment, particularly in the high noise regime. Here, we find that our confidence estimates can be made directly by examining the statistics of the distribution of ensemble points in the coefficient space for a particular signal class. Simply, we aim at calculating the statistical probability that any new data observation is a member of a particular signal class, by calculating the probability that its coefficent space point lies within the coefficient space distribution defined by a given class. Since these distributions are typically non-Gaussian, sophisticated non-parameteric methods will likely provide the best estimators. At present, we only define a simple-minded statistic, which can yield rough preliminary results. We do this by assuming a Gaussian form for the coefficient space distribution, and then calculating the mean and standard deviation for the each particular signal class. An arbitrary point in coefficient space can then be assigned a probability of belonging to a particular distribution by calculating its distance from the distribution mean, normalized in units of its standard deviations.

As we discussed in Sect. 2, we define a criterion for choosing coefficients which is based on the coefficient ensemble dustributions. We calculate the mean, $\langle c \rangle$, and the standard deviation σ_c in the usual way, for each coefficient separately of a given distribution. One can show that for pure noise, we must ideally have $\langle c \rangle \approx 0$ and $|\langle c \rangle|/\sigma_c \ll 1$. The latter property can be used to define a criteria for "significance" of any coefficient value: we consider that any coefficient for which $\langle c \rangle/\sigma \geq 1$ is statistically significant, the contrary implying that its value statistically includes zero. Hence, we may be able to reduce the number of coefficients to be compared during a particular classification application, thereby reducing the probabilities of "false alarms" due to additional axes which contain no information.

For determining whether an arbitrary data observation is a member of a particular class, we can similarly measure the distance between the coefficient space point and the mean of the given signal class distribution, using the significant coefficients. Hence, we can write the distance between centers of clouds from signal 1 and signal 2

$$r_c = \sqrt{\sum (\langle c^{(1)} \rangle_i - \langle c^{(2)} \rangle_i)^2} \,, \tag{4}$$

which is normalized by the sum of weighted variances :

$$\sigma_\Sigma = \sum q_i \sigma_i^{(1)} + \sum q_i \sigma_i^{(2)} \,, \tag{5}$$

where summation is done over significant coefficients and weights are determined by the projection

$$q_i = \frac{|\langle c^{(1)} \rangle_i - \langle c^{(2)} \rangle_i|}{r_c} \,. \tag{6}$$

Since we approximate using Gaussian statistics, this measure can be directly translated into probabilities that a given observation belongs to the given class. It is easy to see that, for example, when r_c/σ_Σ is greater than 1, the probability of a false alarm (when coefficients from a particular observation of Signal 1 are found inside the distribution of the coefficients from Signal 2) in this case is less than 5%. Correspondingly, if $r_c/\sigma_\Sigma \geq 2$, then the probability of the false alarm does not exceed 1%. Similar probability measures to these are quoted in the next section to indicate detection/classification performance. We note that utilization of more sophisticated statistical measures will significantly improve performance in marginal cases, since one profits from the significantly non-Gaussian nature of the distributions (i.e., the distributions are typically tightly grouped).

When analyzing real-world data sets, the guidelines outlined above provide some useful rules-of-thumb for estimating the proper algorithm parameters to provide generally satisfactory detection/classification performance. Of course, for a particular application, these parameters can be tuned to the particular times scales and other nuances of the problem, and a trial-and-error procedure used, typically yielding corresponding increases in performance. In the remainder of this paper, we will use these ideas to present some numerical results from the analysis of acoustic data derived from a real-world application.

4 An Example of a Real-World Application

One of the most important and widespread applications of detection/classification schemes is to acoustic (SONAR) data analysis, where typically most of the non-ideal characteristics of real data (high noise level, short or non-stationary signals, clutter, etc.) are constantly evident. One of the most important questions here is the separation of "biological" signals (those derived from living organisms) from those of man-made devices, which can be quite a difficult problem. The data we analyze here consists of actual open ocean SONAR recordings originally digitized from magnetic tape, consisting of long stretches of background noise punctuated by occasional events consisting of various biological acoustic transients. The data was generously supplied by T. Luginbuhl and M. Gouzie of NUWC, New London. This data was digitized at a sampling rate of 25 kHz, and the signal-to-noise ratio of the transients above background is typically only a few dB. Samples of time series of the data are shown in Fig. 7, where we have included segments consisting of pure background noise, porpoise "whistles", and whale "cries". Event location of the biological signals can usually be determined from amplitude variation, and hence detection is not a major difficulty, so that classification becomes the primary issue. An expanded time domain scale often shows some pulse structure, but not significantly so. Spectral analysis of the pulse regions is also difficult due to the low SNR and non-stationarity.

The principal question we will ask is this: can we construct a classification scheme which can determine the difference (in a statistical sense) between the noise, the porpoise "whistles", and whale "cries"? This data is a particularly difficult test, since the SNR is relatively low, and the SOI are quite non-stationary, are not easily localized, and apparently vary in length and structure. Indeed, there is no reason to assume that such biological signals can be modeled as a low-dimensional dynamical system at all, or hence that this particular method will prove sensitive to any information beyond linear. Although there is some precedent to indicate that biologic vocalizations can

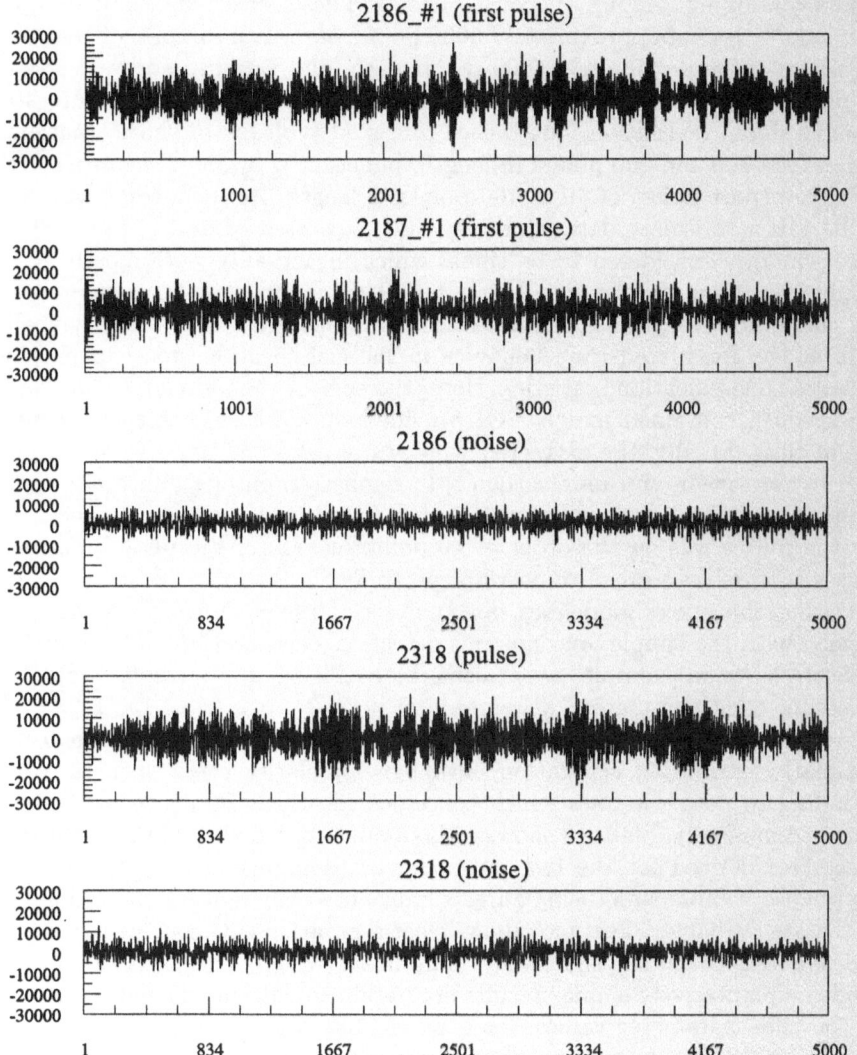

Fig. 7 The samples of whale "cries" (2186 and 2187), porpoise "whistles" (2318) and noisy background from 2186 and 2318 data files.

be modeled as a dynamical system [e.g. Herzel et al., 1994], whether or not this dynamical structure is repeatable from individual to individual (or even from "word" to "word" of a particular individual) is an open question. Here, we proceed in a "black box" fashion, and simply perform the analysis without rigorous justification.

The data files themselves consisted of rather long sequences (at least $3 \cdot 10^5$ points) of background signal, from which were manually extracted several segments containing samples of acoustic events. These consisted of four samples of whale "cries" of approximately 5000 points each, two samples of porpoise "whistles" of approximately 6000 points each, and various nearby segments of pure background noise. Each data segment did not contain a pure stationary signal realization, but rather consisted typically of short transients of between 100 and 200 points in length, punctuated by short segments with no signal (pure noise) of 40 to 100 points in length. As such, some care had to be taken to isolate data windows which consisted of mostly SOI, otherwise windows considered to be signal which in actuality were mostly noise would give erroneous results. In an actual detection/classification processing chain, however, this is not necessarily an issue, since one would typically look at the integrated time behavior in the classification space to provide detection/classification statistics. Here, the sum of the individual data windows which contained mostly SOI provided the data ensemble for a given signal class. Finally, the extracted windows of data and noise were normalized to zero mean and unit variance, to remove amplitude information and standardize the model representation. We found that the typical time scale for the pulses was on the order of 40 points/characteristic period, and the pulses themselves seemed to overlap spectrally.

Using the above guidelines, an analysis of the SONAR data was performed with the simple, autonomous detection/classification code described in Sect. 2. As an example, we present the coefficient space results when the following code parameters were used: $D = 3$, $P = 2$, $\tau = 12$, $L_{sm} = 8$, $L_w = 300$, and $L_s = 150$. These parameters provide a good (although not optimal) distribution separation, although for clarity some outliers corresponding to data windows which contained mostly noise components were removed manually. Figure 8 shows a three-dimensional view of the coefficient ensembles derived for the two pulse classes (dolphin and whale), and the pure noise components, using three significant coefficients of the space (up to ten are available). This plot shows clear separation between the two signal types, and between the signals and pure noise (which is clustered about the origin). Qualitatively similar results are obtained using significant variations in the code parameter values, for example $150 \leq L_w \leq 600$, $7 \leq \tau \leq 15$, $4 \leq L_{sm} \leq 10$, hence the general performance is stable.

A quantitative measure of the classification performance is obtained using the simple statistical arguments discussed above. In Table 1, we summarize the statistical measures $\langle c \rangle / \sigma$ and r_c / σ_Σ, indicating separation from noise (detection) and separation from other signal types (classification), computed for the various data samples analyzed. As the table indicates, the signal distributions are typically separated by 2 to 4 standard deviations, indicating a high confidence level as estimated by the Gaussian approximation. As indicated above, one can roughly estimate Probability-of-Detect vs. probability-of-false-alarm (P_d/P_f) using the Gaussian arguments. In reality, the ensemble

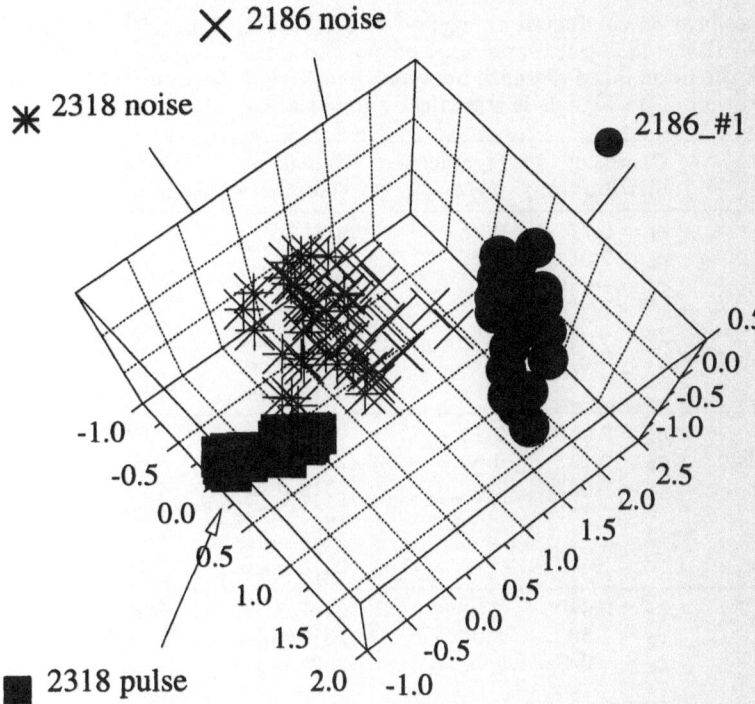

Fig. 8 3-dimensional coefficient space based on linear coefficients $c_{1,2,3}$. In this example we used data from 2186_1 (first pulse) and 2318-pulse and noise was taken from both files, see Table I for the complete numerical presentation of the detection/classification scheme.

distributions are more tightly grouped than Gaussian, hence non-parametric methods likely result in even higher confidence levels.

One interesting point from the analysis is that the ensembles from the different whale pulses nearly overlap. At present, it is not clear whether these separate pulses came from the same whale emitting the same "word", the same whale emitting different "words", or different whales emitting the same or different "words". If the first case holds, then this would imply at least that the classifier is stable in its performance. If the second case holds, then it would imply that this detection/classification method may be capturing some fundamental property (nonlinear correlation) of a given single whale's vocalization cavity, which could be an interesting result. If the third case holds, then the same could be said for the vocalization cavities of the class of whales in general, and this result could be remarkable. Further analysis is underway at present to understand this result more fully.

It should also be noted that this method is numerically quite efficient. The bulk of the computational work (for the analysis of a single window) consists of the least-squares solution of the particular matrix equation, which

Table 1. Classification conducted on whale "cries" (2186_1, 2 and 2187_1, 2) and porpoise "whistles" (2318-pulse and 2324-pulse) and noise (from 2186 and 2318 data files). If the normalized distance between clouds (right column) is more than 1, then separation of the signals is statistically significant.

Signal	Center of distribution	Significance $\langle c \rangle / \sigma$	Signal to compare with	r_c / σ_Σ
2186_1 first pulse	$c_1 = 0.77$	**2.08**	2186_2	0.16
	$c_2 = 2.01$	**6.25**	2187_1	0.44
	$c_3 = -0.33$	**1.1**	2187_2	0.51
			2318-pulse	4.01
			2324-pulse	3.00
			2186-noise	1.61
			2318-noise	1.94
2186_2 second pulse	$c_1 = 0.66$	**2.13**	2187_1	0.64
	$c_2 = 2.13$	**6.87**	2187_2	0.57
	$c_3 = -0.28$	**1.04**	2318-pulse	4.54
			2324-pulse	3.53
			2186-noise	1.86
			2318-noise	2.18
2187_1 first pulse	$c_1 = 0.63$	**2.03**	2187_2	0.33
	$c_2 = 1.58$	**3.06**	2318-pulse	1.28
	$c_3 = -0.15$	0.70	2324-pulse	2.02
			2186-noise	1.02
			2318-noise	1.36
2187_2 second pulse	$c_1 = 0.64$	**2.28**	2318-pulse	3.26
	$c_2 = 1.78$	**4.14**	2324-pulse	2.53
	$c_3 = 0.096$	0.38	2186-noise	1.28
			2318-noise	1.63
2318 pulse	$c_1 = 0.45$	**2.65**	2324-pulse	0.92
	$c_2 = -0.64$	**2.67**	2186-noise	2.26
	$c_3 = -0.13$	0.60	2318-noise	1.50
2324 pulse	$c_1 = 0.17$	**1.76**	2186-noise	1.15
	$c_2 = -0.31$	**3.87**	2318-noise	1.40
	$c_3 = 0.07$	0.70		
2186-noise			2318-noise	0.50

essentially consists of the SVD of a matrix which is typically of size 10 by 300. Overall computation times are multiplied by the number of windows actually analyzed, which may typically be 50 to 100. For these typical numbers, total run time is usually less than 1 minute on an IBM 486DX2-66 class PC. Hence, hardcoding of such an algorithm on a dedicated CPU, which would speed up the computation time significantly, could easily make the technique realizable for real-time analysis.

5 Summary and Future Directions

In this contribution, we have presented an algorithmic scheme and an operational approach for the classification of complex, noisy data, which is robust to many of the non-ideal conditions typically imposed in real data analysis situations. This scheme is interesting from a signal processing point of view, since the model class for the signal structure assumes a dynamical nature for the underlying signal generator, and hence seeks to incorporate higher dimensionality and nonlinearity in the system model in a natural way. From the point of view of nonlinear dynamics, the scheme is perhaps interesting in that the proper assumptions about operational performance allow us to relax the restrictions of global modeling methods, and use estimated global models as a relative measure of linear and nonlinear structure in data, and also to look for determinism as a measure of structure. It also allows us to estimate classification probabilities in a natural way, via the coefficient space representation of model coefficient structure.

In addition to the discussion above, it should be noted that only the simplest algorithm has been discussed here, and a variety of fundamental and technical improvements can be implemented. Different choice of basis set [e.g., rational polynomials; see Kadtke and Brush, 1993], general non-autonomous models, modification of the least-squares problem, and alternate normalization methods should be investigated. In terms of data analysis, optimal choice of time scales for non-stationary data, superior windowing techniques, and a more sophisticated statistical analysis are necessary to obtain the best classification results. We are currently developing several improvements, including an adaptive coefficient estimation procedure [Brush and Kadtke, 1993] for real-time analysis, which we plan to present in the future. Finally, we point out that the numerical efficiency of these algorithms make them quite adaptable to real-time signal analyses, via dedicated processing chips.

Acknowledgements

The authors would like to thank M. Inchiosa, A. Bulsara, C.F.Driscoll, C. Swanson, and G. Mayer-Kress for valuable scientific input; T. Luginbuhl and M. Gouzie of NUWC for supplying the SONAR data; and M. Shlesinger for continued support and encouragement. This work was supported in large part by a grant to the RTA Corp. from the Office of Naval Research, Grant # N00014-92-C-0045.   J.B.K. also wishes to acknowledge travel support from the NATO Office of Scientific Affairs, Grant # 900088.

References

Badii, R. and Politi, A. (1986): In *Dimensions and Entropies in Chaotic Systems* (ed. G. Mayer-Kress), Springer, Berlin, Heidelberg, p. 67.

Brown, R. (1993): Phys. Rev. E **47**(6), 3962.

Brush, J. and Kadtke, J. (1993): *AIP Conf. Proc.* **296**, (ed. R. Katz), AIP Press, 159.

Casdagli, M. (1989): Physica D **35**, 335.

Cawley, R. and Hsu, G. (1991): Phys. Rev. A **46**, 3057.

Cremers, J. and Huebler, A. (1987): Z. Naturforsch. **42a**, 797.

Crutchfield, J. and McNamara, B. (1987): Complex Systems **1**, 417.

Elgar, S. and Chandran, V. (1993): Int. J. Bif. and Chaos **3**(1), 19.

Farmer, J.D., and Sidorowich, J.J. (1987): Phys. Rev. Lett. **59**, 845.

Hammel, S. (1990): Phys. Lett. A **148**, 421.

Hayes, S., Grebogi, C., Ott, E., and Mark, A. (1994): Phys. Rev. Lett. **73**, 1781.

Herzel, H., Berry, D., Titze, I.R., Saleh, M. (1994): J. Speech and Hearing Res. **37**(5), 1008–1019.

Kadtke, J. (1995): Phys. Lett. A, **203**, 196.

Kadtke, J. and Brush, J. (1993): *AIP Conf. Proc.* **296** (ed. R. Katz), AIP Press, 205.

Kadtke, J., Brush, J., and Holzfuss, J. (1993): Int. J. Bif. and Chaos **3**, 607.

Kostelich, E., and Yorke, J.A. (1990): Physica D **41**, 183.

Mindlin, G. and Gilmore, R. (1992): Physica D **58**, 229.

Pecora, L. and Carroll, T. (1990): Phys. Rev. Lett. **64**, 821.

Press, W., et al. (1986): *Numerical Recipes*, Cambridge, p.509.

Salvino, L. and Cawley, R. (1994): Phys. Rev. Lett. **73**(8), 1091.

Takens, F. (1981): In *Dynamical Systems and Turbulence* (ed. D. Rand and L.S. Young), Springer, Berlin, Heidelberg, p.366.

Theiler, J. et al. (1992): Physica D **58**, 77.

Part 3

Dynamical Modeling
and Forecasting Algorithms

Strategy and Algorithms
for Dynamical Forecasting

Oleg L. Anosov, Oleg Ya. Butkovskii and Yurii A. Kravtsov

Abstract

The strategy and algorithms for reconstructing dynamical equations directly from observed time series are discussed. A short review of the research literature is presented along with a technique for reconstructing the autonomous dynamical system in a given class of ordinary differential equations; these consist of a set of first-order differential equations with polynomial nonlinearity. The coefficients of these nonlinear terms are determined by the condition of fitting the best equations to the experimental time series. An important feature of the suggested algorithm is that it is able to restore hidden (not observed) variables from one-dimensional time series. The applications of this technique are illustrated by numerical examples for several dynamical systems, among them the Lorenz and Rossler attractors. The limiting accuracy of the differential equation reconstruction depends strongly on the noise level. As a rule, however, the reconstructed dynamical system manifests predictable behavior on time scales exceeding the linear correlation time of the observed process.

1 Drawing a Distinction Between Randomness and Determinism

The ability to forecast data from a process evolving in time, or, as researchers in natural sciences would say, say a "time series", is very much called for in different areas of human endeavor. In this work we will try to outline some approaches to the problem of dynamical forecasting, that is, forecasting based on certain dynamical laws.

Let us consider a process recorded from a sensing element and represented in Fig. 1.

The first question that perhaps arises as we look at the data is about the nature of the process, that is, whether it is random or deterministic. Even at this early stage of our research we encounter a theoretically profound and practically essential problem of how to distinguish between randomness and determinism. It is not until the nature of the process is identified that we can select a forecasting method. In fact, we have to be satisfied with methods to construct predictive models, since forecasting is always based on a distinct model.

y(t)

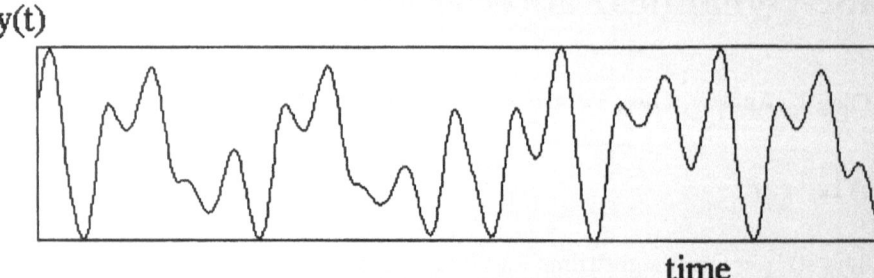

time

Fig. 1 Example of a random-like process of dynamical nature. Such processes exhibit both stochastic (in a sense) and deterministic properties.

If a process is judged to be random, it is only natural to turn to statistical methods of prediction such as originally worked out by Wiener [1950, 1957].

If there are signs or suspicions that the process has a dynamical origin, it is more expedient to try a dynamical forecast based on relevant regularities [attempts to apply such an approach are reported by Breeden and Hübler, 1990; Gouesbet, 1991; Brush and Kadtke, 1992; Gribkov et al., 1994a, 1994b to name but a few]. The difference between the statistical and dynamical forecasts can be truly significant.

A statistical forecast starting from probabilistic and correlation information is usually limited by relatively short intervals of time comparable with the autocorrelation time τ_c of the process. For cases where the dynamical origin is known, it seems logical to use dynamical forecasting based on nonlinear dynamics inverse problems, which stems from ideas of Kolmogorov and Gabor to reconstruct dynamical systems directly from noisy experimental series. Today's state-of-the-art is presented in books such as [Mandrekar and Saleki (Eds.), 1983] and [Ljung, 1987]. By way of example, we can refer to weather forecasting in which dynamical methods based on hydrodynamical equations for air flows are used to advantage [Mason and Treas, 1986]. Admittedly, the dynamical methods also fail to predict a process far into the future. One of the main reasons for this is the local instability inherent in most of the phenomena around us, and governing such processes as dynamical chaos.

A peculiarity of equations yielding dynamical chaos is that they fail to provide reliable long-term forecasts. The point is that errors in initial conditions, along with other hindering factors (noise inevitable in real systems, small inaccuracies occurring in model equations, a slight nonstationarity of a system, etc.), tend to grow exponentially under local instability conditions, so that after some finite time all the researcher's efforts would be in vain.

Taking A to be the typical amplitude of oscillations under study and δ the perturbation in the system occurring due to one or another factor, the limit time of predictable behavior τ_{lim} can be roughly estimated as:

$$\tau_{\lim} = \frac{1}{2\lambda_+} \ln \left(\frac{A}{\delta} \right) , \tag{1}$$

where λ_+ is the maximum Lyapunov exponent giving the perturbation growth rate.

Although the prediction time is limited, as follows from Eq. (1), it can exceed (and often does exceed!) the autocorrelation time τ_c. It is the ratio $\tau_{\lim}/\tau_c$ that characterizes the "gain" provided by the dynamical forecast with respect to the statistical forecast. The main question is how to realize this gain when the dynamical equations governing a system's behavior are not known a priori.

To reconstruct these equations while analyzing a time series is the "inverse" problem of nonlinear dynamics. It is this problem that we will address here. We will show schematically how differential equations are reconstructed directly from experimental data, treat some specific reconstruction algorithms, and demonstrate them with examples.

When using the dynamical approach, elements of statistical analysis and data processing need also to be employed. Moreover, apart from purely statistical (probabilistic) and consistently dynamical approaches, there exist a number of "intermediate" methods, based partly on probabilistic, partly on dynamical algorithms. Without going into such combined approaches, we only mention that the forecasts they provide are of intermediate nature, too.

In choosing between a statistical and dynamical forecasting strategy, the scientist must realize that, with different test methods, one and the same time series may be found at times to be random and at other times deterministic. The problem is that several agreements on how to define "randomness" coexist in modern science, and therefore a process satisfying a randomness test under one agreement can be judged to be deterministic under another.

The question of these different agreements on randomness has been addressed by [Kravtsov, 1989, 1993]. The most widespread agreement is the one underlying the modern set-theoretical approach and the theory of probability. Within this framework, anything having a probabilistic measure is deemed to be random. Alternately, the algorithmical approach identifies randomness with algorithmical complexity.

Physicists generally hold to the "set-theoretical" agreement, while entertaining in some cases their own views of randomness, such as the criterion of decaying correlations, or the criterion commonly adopted by the experimental physical community, according to which randomness is identified with unpredictability. The relations developing under this agreement have been formalized only recently [Kravtsov, 1989, 1993].

Note that processes like dynamical chaos (predictable for times τ_{pred} and thus seemingly deterministic) appear to be random in the context of many other agreements (by the criterions of algorithmical complexity, decaying correlations, etc.). That is why in tackling the inverse problem of nonlinear dynamics one can completely neglect the results of many randomness tests, if they are useless in reconstructing the governing dynamical equations. Of course, the most important predictability test is correlation between what is observed and what was predicted. A suitable qualitative indicator of such a correlation is the degree of predictability (that is, determinism) which is introduced as a normalized coefficient of correlation between the observed and the predicted [Kravtsov, 1989, 1993]. The typical time in which such a correlation decays is called the predictability time. For dynamical chaos, the predictability time is roughly estimated according to Eq. (1).

2 Assumptions and Prerequisites

To predict using a dynamical approach, it is expedient to narrow the class of equations used to describe a given process. Otherwise the problem of dynamical forecasting becomes as indefinite as it is difficult to solve. Therefore, let us point out a number of restricting assumptions.

First, we exclude from our consideration those spatio-temporal processes whose description requires partial differential equations to be employed. Instead, we will deal with dynamical systems given by ordinary differential equations. Second, we restrict ourselves to autonomous models involving no external forcing. Third, we assume the equations to be stationary. Whenever autonomy and stationarity are not assumed, the reconstruction algorithm gets dramatically complicated, though some ordinary nonstationary systems, e.g. those with a stepwise change of parameters, allow reconstruction algorithms which are not unduly complicated [see, for example, Anosov et al., 1995a,b].

Fourth, in elaborating particular algorithms, we will consider to polynomial approximations of nonlinear functions appearing in differential equations under reconstruction.

Even under these constraints, reconstruction remains an extremely indeterminate and tedious task. For example, when taking the exhaustive search strategy and trying equations from a rather limited model class one-by-one to fit the experimental data, we will see that no modern data processing techniques are powerful enough to meet this challenge. Fortunately, there are some algorithms available for such a purpose that are more effective than the exhaustive search method. Their elaboration was possible due to the following favorable factors.

In the first place, we point out the experimental fact that the majority of nonlinear systems exhibiting chaotic behavior have a relatively low dimension $n \sim 3-5$. The abundance of low-dimensional systems is what obviously makes one believe that the inverse problem of nonlinear dynamics is solvable.

In the second place, there is much evidence that the evolution of dynamical processes, even in widely different systems, generally follows one of a limited set of scenarios. The number of typical scenarios is relatively small –

not more than ten for low-dimensional systems. This is where the dynamical isomorphism principle reveals itself. It permits the class of tested systems of differential equations to be further confined.

Furthermore, one must be aware of the possibility of reconstructing the variables hidden from observation. Such a possibility only seems to be realizable in highly nonlinear systems where different variables are strongly interconnected, each bearing the information about the whole system.

Another favorable factor, even though unable to confine the class of differential equations, allows a considerable saving of processing time. This is that increasing the length of the time series under examination beyond a limiting value τ_{lim} does not necessarily lead to a better accuracy of the coefficients of an equation under reconstruction, but invariably extends the calculation time. As a result, one can reduce the data processing time without sacrificing the accuracy of the model equation. Unfortunately, it is impossible to determine the limiting time τ_{lim} a priori, since it is to be defined in the course of data processing. One can only guess that the lower limit of τ_{lim} is the time of prediction carried several time steps into the future.

Now let us consider, following the earlier works [Gribkov et al., 1994a, 1994b], some specific procedures whereby predictive differential equations can be obtained. We will give most attention to revealing the hidden (unobservable) variables.

3 Techniques for Reconstructing Dynamical Equations from Time Series

3.1 Filtration

Let us consider the values of an observable process $x(t)$ at discrete times t_k : $x_k \equiv x(t_k)$ with the understanding that here and time indices are indicated by subscripts. We take the sampling interval $\Delta t = t_{k+1} - t_k$ small enough to satisfy the condition $\omega_p \Delta t \ll 1$, where ω_p is the frequency corresponding to the spectrum peak. The number of samples is normally several hundred to several thousand.

Given a high-frequency ($\omega_p \ll \omega$) noise, it is worthwhile "pre-filtering" the process $x(t)$ under consideration, so that it could be differentiated a necessary, though limited, number of times.

As a result of filtering the observed process we obtain a fairly smooth "dynamical" component $X(t)$ and the remainder $\xi(t)$ which according to convention, will be called noise:

$$x(t) = X(t) + \xi(t) . \tag{2}$$

The conventionality involved in dividing $x(t)$ into the "dynamical" and "noisy" components stems from the fact that the "noisy" component invariably incorporates the high-frequency portion of the dynamical process, while the "dynamical" portion contains the low-frequency noise component. This is where the scientist must demonstrate his sense of proportion, for too deep a filtering can cause unacceptable distortions of the dynamics, while a superficial filtration fails to provide the required degree of differentiation.

The high-frequency noise filtration is effected in different ways. [Gribkov et al., 1994a,b] recommend a multiple (up to 50–80 times) filtering using the least-squares algorithm [Teodorescu, 1989] with a weight coefficient of about 0.5, sometimes in conjunction with the method described by [Grassberger et al., 1992].

Figure 2 demonstrates an example due to [Gribkov et al., 1994b] of a noise-polluted process resulting from the addition of a 15% short-correlated noise to the originally Lorenz process shown in Fig. 2(a). The result of filtration by the said algorithm is very close to that of the original process (see Fig. 2(c), the smooth curve $X(t)$).

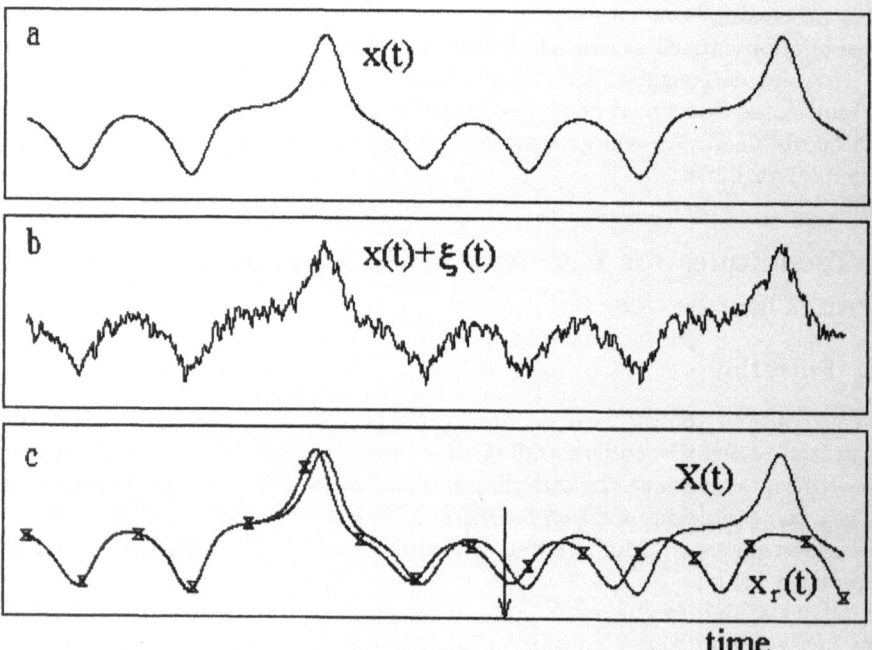

Fig. 2 The effect of noise $\xi(t)$ added to the Lorenz system process is that (a) becomes (b). In (c) a filtered process $X(t)$ is shown by the unmarked curve, whereas the reconstructed process $x_r(t)$ is marked. The predictability time designated by the arrow corresponds to a 15% divergence of the curves [Gribkov et al., 1994b].

3.2 Determining the Dimension and Constructing a New Basis

To find out about the dynamical nature of the time series $X(t)$ one usually employs Takens delayed variables [Takens, 1981] $X_1(t) \equiv X(t)$, $X_2(t) \equiv X(t + \tau), \ldots, X_n(t) \equiv X(t + (n-1)\tau)$, which can be pieced together in an n-component vector $\boldsymbol{X} = \{X_1, \ldots, X_n\}$. Here the time delay τ is chosen close to the sampling rate Δt of the observed time series. Using conventional procedures [Grassberger and Procaccia, 1983], one can determine the dimension of the embedding space.

Let us introduce the state vector $\boldsymbol{X}^k$ at instant t_k with components: $X_1^k \equiv X(t_k)$, $X_2^k \equiv X(t_k + \tau), \ldots, X_n^k \equiv X(t_k + (n-1)\tau)$, $k = 1, \ldots, N$ where N is the total number of time series points and n is the initial embedding dimension. According to [Broomhead and King, 1986], an orthogonal basis can be constructed using eigenvectors S_i of the correlation matrix:

$$C_{kl} = \sum_{i=1}^{n} X_i^k X_i^l, \tag{3}$$

with any state vector $\boldsymbol{X}^k$ expandable in the orthonormal vector set:

$$\boldsymbol{X}^k = \sum_{i=1}^{n} \tilde{X}_i^k \boldsymbol{S}_i. \tag{4}$$

The coefficients of the $\boldsymbol{X}^k$ expansion are scalar products:

$$\tilde{X}_i^k = (\boldsymbol{X}^k, \boldsymbol{S}_i). \tag{5}$$

Expression (5) allows new variables $\tilde{X}_i$ to be determined. The number of "principal" basis set vectors, to where the state vector $\boldsymbol{X}^k$ is projected, is bounded by the value $n_0 < n$, which we take to be the system's dimension. The variables $\tilde{X}_i$ for $i > n_0$ become small enough for the number of basis vectors to be actually reduced to n_0.

3.3 Choosing an Optimal Basis Set

Taking a small τ the components of the state vector $\boldsymbol{X}^k$ can be expanded into a Taylor series:

$$X_i^k = X(t_k + (i-1)\tau) = \sum_{m=0}^{n_0} \partial_t^m X(t_k) \left[\frac{(j-1)\tau}{m!} \right]^m + O(\tau^{n_0+1}), \tag{6}$$

where $\partial_t^m X(t_k)$ denotes the mth time derivative of $X(t_k)$. According to (6), each of $\tilde{X}_i^k$ variables is expanded into an nonorthogonal set of functions: $X_1^k, \ldots, \partial_t^m X_1^k, \ldots$. At $\tau \to 0$ the contribution of the high order derivatives gets negligibly small, and in (6) it becomes possible to confine ourselves to the dimension n_0 of the system.

Using the inverse linear change of variables we can express the process X_1^k and its first derivatives as a combination of variables $\tilde{X}_i^k$. In the third-order system ($n_0 = 3$) we obtain with satisfactory accuracy:

$$X_1^k = a\tilde{X}_1^k,$$
$$\delta_t X_1^k = b\tilde{X}_2^k,$$
$$\delta_t^2 X_1^k = c\tilde{X}_2^k + d\tilde{X}_3^k.$$

(7)

What follows is the technique for reconstruction of ordinary differential equations using the example of a three-dimensional system [Gribkov et al., 1994a,b].

3.4 Reconstruction of Differential Equations

Using relations (7), one can introduce the system of equations of the form:

$$\dot{X}_1 = \alpha_1 X_2,$$
$$\dot{X}_2 = \alpha_2 X_3,$$
$$\dot{X}_3 = \frac{F(X_1, X_2, X_3)}{f(X_1, X_2, X_3)}.$$

(8)

Here X_j^k are renamed simply as X_j ($j = 1, 2, 3$), the dot denoting time differentiation, α_1, α_2 being constants, and f one of monomials of the form 1, X_i, X_j, $X_i X_j$ ($i, j = 1, 2, 3$). F is a polynomial with indeterminate coefficients that can be represented as a sum of all possible monomials G_j composed of the variables X_1, X_2, X_3 :

$$F(X_1, X_2, X_3) = \sum A_j G_j(X_1, X_2, X_3).$$

(9)

The task of seeking the coefficients A_j can be reduced to searching for the minimum of the functional

$$\phi = \sum_{k=1}^{N} [f^k \dot{X}_3^k - F^k]^2,$$

(10)

where f^k and F^k are values of the functions f and F at instant t_k. All the values of the variables $X_{1,2,3}^k$ from the initial time series in (10) must be recalculated in view of the performed substitution.

The minimum of the functional ϕ is achieved under the following condition:

$$g = GA,$$

(11)

where $g = \{g^k\}$, $g^k = f^k \dot{X}_3^k$ and $A = \{A_j\}$ is a vector of coefficients A_j to be determined. The "nonlinearity matrix" G with elements G_j^k is a set of monomials G_j, their values at $t = t_k$, $k = 1, \ldots, N$ being:

$$\tilde{G} = \begin{pmatrix} 1 & x_1^1 & x_2^1 & x_3^1 & (x_1^1)^2 & x_1^1 x_2^1 & \cdots \\ \vdots & & & & & & \\ 1 & x_1^N & x_2^N & x_3^N & (x_1^N)^2 & x_1^N x_2^N & \cdots \end{pmatrix}$$

The system of equations (11) is formally solved as

$$A = (G^T G)^{-1} G^T g. \tag{12}$$

With these coefficient values one obtains the reconstructed dynamical equations (8) which now have a predictive power. However, this is not where the reconstruction procedure ends.

3.5 Selection of Coefficients

We can reduce the complexity of equations (8) with the coefficients (12) by discarding small coefficients, in the hope that this will not adversely affect the prediction power. More often than not this method proves efficient, but sometimes discarding of small coefficients can drastically alter the solution of Eq. (8).

To avoid such an effect, one can apply a more complex but more reliable search algorithm, by which only those coefficients A_j are preserved that exhibit small variations within a time series. These variations are estimated by the value of diagonal elements of the matrix $G^T G^{-1}$. Thus, we only preserve coefficients fitting the inequality $A_i \gg \delta A_i$ and omit (assume equal to zero) those coefficients that show unacceptably large variations $\delta A_i \gg A_i$.

3.6 Providing Global Stability

According to the papers of [Gribkov et al., 1994a,b] a reconstructed system is more tolerant of small disturbances once it has the property of dissipativeness. In a dissipative system, the phase volume element gets smaller with time, corresponding to the negative value of $\operatorname{div} V < 0$, where $V_i = \dot{X}_i$.

In the specific case of the three-dimensional system (8) we have

$$\frac{\partial V_1}{\partial X_1} = \frac{\partial V_2}{\partial X_2} = 0, \qquad \frac{\partial V_3}{\partial X_3} = \frac{\partial}{\partial X_3}\left[\frac{F}{f}\right].$$

Here the condition of compressibility of the phase volume takes the form:

$$\operatorname{div} V = \frac{\delta V_3}{\delta X_3} = \frac{\delta}{\delta X_3}\left[\frac{F}{f}\right] < 0.$$

In practice, it suffices to declare the mean value of the divergence to be negative:

$$\langle \operatorname{div} V \rangle = \frac{1}{M}\sum_{i=1}^{M} \frac{\partial}{\partial X_3}\left(\frac{F(X_1, X_2, X_3)}{f(X_1, X_2, X_3)}\right) < 0. \tag{13}$$

Here summation is extended over M points uniformly distributed within a sufficiently small volume V of the phase space (x_1, x_2, x_3).

By imposing the dissipativeness condition one can drop dubious, that is, small or highly fluctuating, coefficients A_i, provided the condition (13) is met for the reduced model. This imparts some additional stability to the reduced model, even if, perhaps, sacrificing local predictability.

4 Examples

4.1 Noise-Polluted Lorenz System

There have been a number of reported attempts to make a consistent dynamical forecast based on differential equations reconstructed from experimental time series. [Breeden and Hübler, 1990] treated an idealized case of a noiseless system, reconstructing equations from experimental data with three hidden variables. [Gouesbet, 1991] implemented an even more general reconstruction scheme than the system of equations (8): he represented the function f as a series with undetermined coefficients, similar to the function F. Reconstruction of equations by Gouesbet was carried out in the absence of noise, from experimental data with three variables. The experimental data with three variables were also used by [Brush and Kadtke, 1992].

A more complex task of reconstructing equations from a single-component noisy time series was solved by [Gribkov et al., 1994a,b] and presented in the following. The initial realization of the variable $x(t)$ shown in Fig. 2(a) fits the Lorenz system:

$$\dot{x} = -\sigma(x - y),$$
$$\dot{y} = rx - y - xz, \tag{14}$$
$$\dot{z} = -bz + xy,$$

where $\sigma = 10$, $r = 45$, $b = 8/3$. The $x(t)$ realization encompasses 2500 points spaced at 0.01. Each realization $x(t)$ was added up with a wideband artificial noise of the form

$$\epsilon(t) = A_0 \delta(t) \cos p(t),$$

where $\delta(t)$ and $p(t)$ are random values uniformly distributed over the intervals 0 to 1 and 0 to 2π, respectively. The amplitude of noise A_0 was 15% of that of the signal. The noisy Lorenz process is shown in Fig. 2(b).

After filtration, the noisy process $x + \epsilon$ was smoothed to $X(t)$, the process represented in Fig. 2(c) by the unmarked curve. The marked curve is the result of restorating equations. Reconstruction of governing equations was performed following the above procedure. After returning to the initial variables, the reconstructed system of equations appeared to very much resemble the Lorenz system, Eq. (14). At 1% noise, the coefficients of the nonlinear

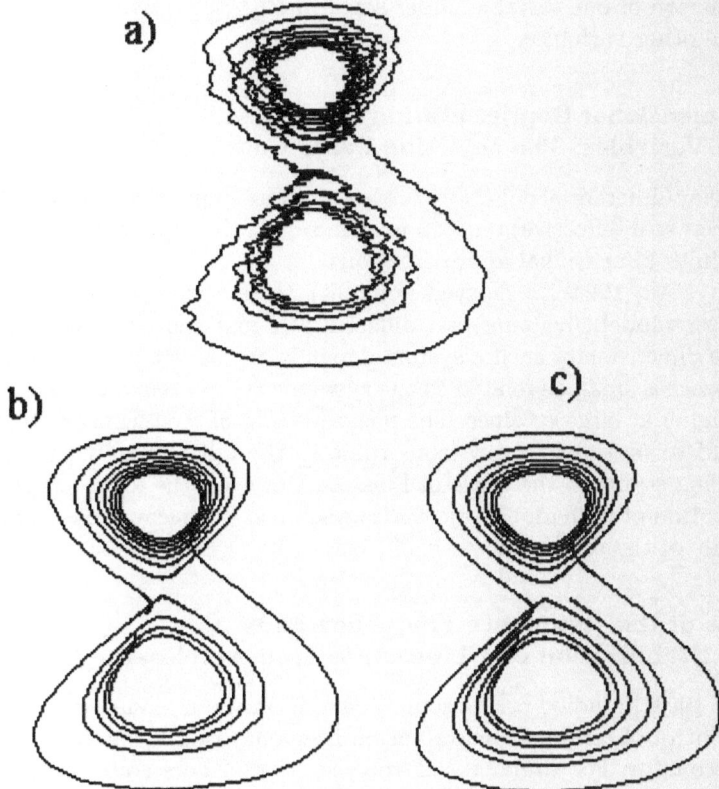

Fig. 3 A noise-polluted phase portrait of the Lorenz system (**a**); the same portrait after filtration (**b**); a reconstructed phase portrait (**c**).

terms missing in the initial system (14) were on the average smaller than 0.02. Of the same order were relative errors of the reconstructed coefficients as against the initial coefficients. At 40% noise, the coefficient errors grew to 5–10%.

Figure 3 presents phase trajectories for the Lorenz system: the noisy trajectory, at 4% noise level (Fig. 3(a)), the trajectory after filtering (Fig. 3(b)) and the trajectory corresponding to the reconstucted differential equations (Fig. 3(c)). From these plots one can make judgements about the effectiveness of the filtration and some qualitative conclusions concerning the identity between the initial and the reconstructed phase trajectories.

The curves in Fig. 4 suggest that reconstruction of the hidden variables y and z has been effective: the reconstructed values y_r and z_r follow the initial values y and z throughout the predictability time common to all three variables. This is another proof of a strong interconnection between the variables

x, y and z: deviation of one variable under some forcing immediately causes departure of the other variables.

4.2 Finite-Dimensional Representation for Systems with Delayed Variables: The Injection Laser Case

Formally, the delay-differential equations have an infinite number of degrees of freedom. But their real (effective) dimension is finite, which in fact encourages the search for finite-dimensional approximations.

In [Gribkov et al., 1994b] a delayed nonlinear system is represented by an injection laser model involving two differential equations with delayed arguments. The dimension of such a system depends on the delay τ, varying from $D = 2$ when τ is small as related to the system's typical relaxation time to $D = 4$ and higher at large τ values. The phase pattern of the reconstructed system appeared to be topologically isomorphic to the initial one, with the system itself being stable in the sense of Poisson. However, the local (short-term) representation of individual trajectories was unstable because of errors in the magnitude of the reconstructed coefficients.

4.3 Dynamics of the Resonance Frequency and Temperature Stabilization of a Linear Electron Accelerator Unit

[Gribkov et al., 1994c] studied oscillations occurring in the nonlinear system of automatic control of temperature and eigen frequency of a linear electron accelerator's resonator. For simplicity of analysis, the authors embarked on a strong smoothing of the process's short-term component. The result was reconstruction of the differential equations of the system, but the coefficients of the system of equations were not very stable, and the predictability time hardly exceeded the autocorrelation time τ_c. The most likely reason for our failure to make a better reconstruction has to do with an over-filtration that had reduced the system's dimension and thus affected the reconstruction accuracy.

4.4 Revealing Nonstationarities in a Nonlinear System

All the systems discussed thus far met the stationarity requirement. There are, however, some algorithms designed to reconstruct differential equations in stationary conditions, which are also useful for revealing sudden nonstationarities like abrupt changes of parameters. Such changes can be revealed in at least two ways. First, one can localize the abrupt change of the parameter by tracing the behavior of coefficients of a system of equations under reconstruction. The accuracy of localization in this case will depend on the length of the processed segment.

Second, one can compare (discriminate) the averaged values in two neighboring time windows. It is this method of nonlinear discriminative analysis

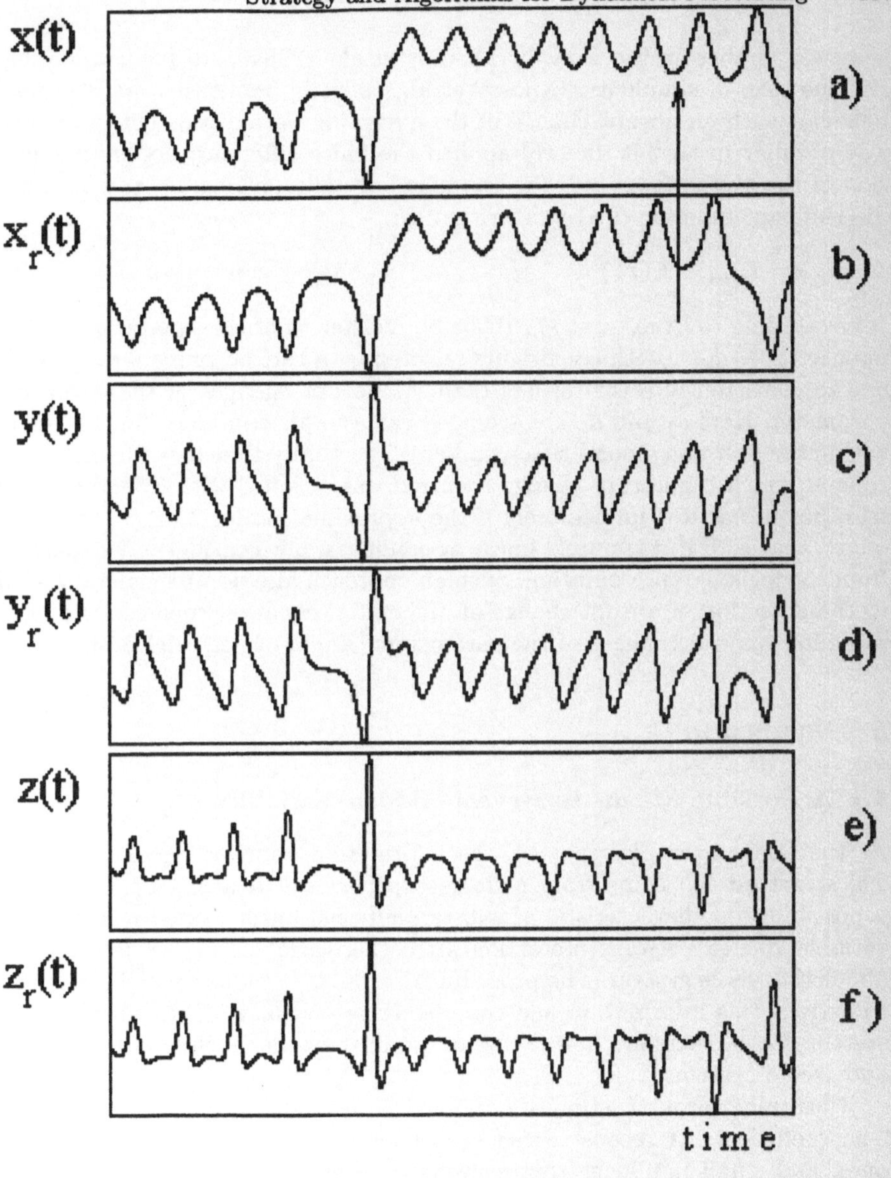

Fig. 4 Example of reconstruction of the hidden variables $y(t)$ and $z(t)$ from a single observable variable $x(t)$ in Lorenz system. The curves (**a**), (**c**), and (**e**) denote initial variables $x(t)$, $y(t)$ and $z(t)$, respectively, while the curves (**b**), (**d**), and (**f**) represent the reconstructed variables $x_r(t)$, $y_r(t)$, and $z_r(t)$. The predictability time is marked in (**a**) by a vertical arrow.

that was applied in the work by [Anosov et al., 1995a,b] to point mapping. For the sake of simplicity, [Anosov et al., 1995a,b] analyzed a logistic map behavior with an abrupt change of the governing parameter. This approach was peculiar in that it directly applied the finite-difference nonlinear equation of the system $x_{n+1} = F(x_n)$ as function under averaging, and thus the discriminant function $d(x)$ was written as:

$$d(x) = x_{n+1} - F(x_n).$$

According to [Anosov et al., 1995a,b], registering of the squares of the difference $(\bar{d}_1 - \bar{d}_2)^2$, with coefficients selected by a certain procedure, allowed one to consistently reveal of less than 1% abrupt changes of the governing parameter. Here $\bar{d}_1$ and $\bar{d}_2$ are averaged values of discriminant function $d(x)$ within the first and second time-windows. The ability to detect the instant of time at which the abrupt change occurred was retained with 1% noise overlaid, but it was lost immediately if the approximation for $F(x_n)$ was inadequate, that is, if it was purely linear or nonlinear, but considerably departing from the logistic representation. Such an approach also admits generalization to the detection of abrupt change of differential equation coefficients and to revealing smooth changes of the parameters [Anosov et al., 1995a,b].

5 Discussion

5.1 A Possible Means to Reveal Hidden Variables

As the above examples suggest, the information about a system's dynamical structure can come from a single experimental relationship $x(t)$. Such a possibility is characteristic of substantially nonlinear systems in which a variable, due to a strong interaction with other variables, carries information about the whole system. The procedures we have developed make it possible to extract this information and completely reconstruct a three-dimensional system. Researchers are now trying to establish similar techniques for $D = 4$ and $D = 5$ systems.

Naturally, reconstruction of a dynamical system is always in some error. The coefficients of reconstructed equations are not identical to the initial ones, and small nonlinear terms absent in the initial equation can appear after reconstruction. Over large time intervals, the "newly acquired" nonlinearities can in principle alter a system's phase portrait so that it becomes unrecognizable. In many instances their occurrence causes the system of differential equations to lose the property of global stability, with the phase trajectory violating the limited phase volume bound. Therefore, in order to rule out the possibility of nonlinearities causing a dynamical system to lose its stability, one can apply the selection technique based on calculation of the mean divergency of a phase space for a certain phase space region containing the phase trajectory under study. The compression of the phase volume under this technique ensures global stability (finite dimension of the attractor).

5.2 Comparison: The Method of Nonlinear Autoregression

An autoregression prediction is made using formulas of the type:

$$x_{n+1} = g(x_n, \ldots, x_{n-1}),\tag{15}$$

where $g(x_n, \ldots, x_{n-1})$ is a polynomial in $\{ x_n, \ldots, x_{n-1}\}$ variables to the mth degree. Polynomial coefficients are usually selected on the basis of best fitting the experimental data. The autoregressive method of prediction is widely used in practice. Let us take a brief look at the effectiveness of the dynamical and autoregressive methods.

Roughly, at fairly small noise levels (not more than 5% in amplitude), the time for which a dynamical system can be predicted by a dynamical algorithm τ_{dyn} is usually greater than the time τ_{AR}^{nl} corresponding to the algorithm of nonlinear autoregression, which in turn is two to three times greater than predictability time $\tau_{\mathrm{AR}}^{\mathrm{lin}}$ obtained by the linear autoregression algorithm. The latter does not go beyond the limit τ_c of the autocorrelation of the process under review. Thus, the predictability times can be written as a set of inequalities:

$$\tau_{\mathrm{dyn}} > \tau_{\mathrm{AR}}^{nl} > \tau_{\mathrm{AR}}^{\mathrm{lin}} \sim \tau_c.$$

At high noise levels (more than 20%) both the dynamical and autoregression predictability times go down. In this case τ_{dyn} is, at best, twice as long as τ_{AR}^{nl}.

With noise growing to 40% or more, the difference between τ_{dyn} and τ_{AR}^{nl} completely disappears: both become equal to the autoregression time. All this suggests that dynamical forecasting offers some advantages over autoregression at low noise levels, but at higher noise these advantages vanish. Nonlinear autoregression yields better results than its linear counterpart, but its predictive capacity is smaller than that of dynamical predictive algorithms.

Without going into details, we state that the nonlinear autoregressive method is not very sensitive to the system's dynamics. At least, a nonlinear autoregression by the algorithm (15) may yield a predictability time which is practically the same for both a noisy and a noiseless after-filtration process, while the dynamical algorithm demonstrates a very high sensitivity to the remaining after-filtration noise. The lessened sensitivity of the autoregression methods to the structure of a dynamical system is due to excessive information contained in delayed variables and to the ensuing additional nonlinearities absent in a real dynamical system. We think that the capabilities of the nonlinear autoregression algorithm can be enhanced by optimizing the choice of the steps l (order of regression), shortening the time τ between the samples, and filtering the initial process. Actually, what this means is imparting properties peculiar to dynamical prediction to a nonlinear autoregression algorithm.

5.3 Possible Application Areas for Dynamical Algorithms

The technique for reconstructing dynamical equations from experimental time series has a wide application potential. Among the possible fields of use are medicine, economics, and information processing and transfer.

In medicine, it can be used for revealing dynamical characteristics of heart and brain activity. The coefficients of reconstructed equations can serve as diagnostic indicators of a new type. The technique is promising as a groundwork for new methods of diagnosing heart muscle ischemia and dystrophy and as a way to facilitate the search for "dynamical" precursors of the arhythmia.

In economics, the use of the dynamical prediction methods opens up a radically new possibility for going beyond the limitations inherent in the standard autoregression methods of prediction of the type (15). A dynamical prediction is far from an all-purpose tool in economics. It is only applicable to low-dimensional, close to autonomous, systems. Unfortunately, we have no list of such systems at hand, but suppose that macro-economic conglomerates wherein the dynamics of subsystem interaction prevails over the contribution of individual economic subjects are well suited for being analyzed and predicted by the dynamical algorithms. Another possible application in economics is isolation of a "noisy" non-dynamical component. In truth, the notion of noise in economics still has to be defined. In our opinion, the philosphy of handling the inverse problems of nonlinear economics forms the right methodical groundwork for meeting this task. The algorithms under consideration hold much promise for reconstructing the dynamics of chemical, biological, and ecological systems.

References

Anosov, O.L., Butkovskii, O.Ya., Isakevich, V.V., Kravtsov, Yu.A. (1995): Nonstationarity revealed in random-like signals of dynamical nature. Radiotekhnika i Elektronika **40**(2), 255–260 (Russian). [Sov. J. Communication Technology Electronics, 1995, **40** (2)].

Anosov, O.L., Butkovskii, O.Ya. (1995b): A Discriminant procedure for the solution of inverse problem for nonstationary systems. This book.

Breeden, J. and Hübler, A. (1990): Reconstructing equations of motion from experimental data with unobserved variables. Phys. Rev. A **42**(10), 5817–5826.

Broomhead, D.S. and King, G.P. (1986): Extracting qualitative dynamics from experimental data. Physica **20**D(2), 217–236.

Brush, J.S. and Kadtke, J.B. (1992): Nonlinear signal processing using empirical global dynamical equations. Proceedings of the ICASSP-92, San-Francisco, 321–325.

Gouesbet, G. (1991): Reconstruction of the vector fields of continuous dynamical systems from numerical scalar time series. Phys. Rev. A. **43**(10), 5321–5331.

Grassberger, P. and Procaccia, I. (1983): Phys. Rev. Lett. **50**(5), 346–349.

Grassberger, P., Hegger, R., Kantz, H., Schaffrath, C., Schreiber, T. (1992): On noise reduction methods for chaotic data. Preprint WVB 92-13, Phys. Dept., Univ. of Wuppertal, Germany.

Gribkov, D.A., Gribkova, V.V., Kravtsov, Yu.A., Kuznetsov, Yu.I., Rzhanov, A.G. (1994a): Reconstruction of dynamical system structure from time series. Radiotekhnika i Elektronika **39**(2), 241–250 (Russian). [Sov. J. Communication Technology Electronics, 1994, **39**(2)]

Gribkov, D.A., Gribkova, V.V., Kravtsov, Yu.A., Kuznetsov, Yu.I., Rzhanov, A.G. (1994b): Reconstruction of differential equations for autostochastic systems using single variable time series. Zhurnal Tekhnich. Fiziki, **3**, 255–260 (Russian).

Gribkov, D.A., Gribkova, V.V., Kravtsov, Yu.A., Kuznetsov, Yu.I., Rzhanov, A.G., Chepurnov, A.S. (1994c): Modeling of systems of resonant frequency and temperature stabilization for a section of linear electron accelerator based on experimental data. Vestnik MGU, ser. fiz., astron. **35** (1), 96–98 (Russian). [Bulletin Moscow University, ser. Phys., Astron., 1994, **35**(1)].

Kravtsov, Yu.A. (1989): Randomness, determinateness, predictability. Sov. Phys. Uspekhi **32**(5), 434–449.

Kravtsov Yu.A. (1993): Fundamental and practical limits of predictability. In: Limits of Predictability, Yu. A. Kravtsov (Eds.), Springer-Verlag, Berlin, 173–203.

Ljung, L. 1987: System Identification: Theory for the users. Prentice-Hall, Berlin.

Mandrekar, V., Saleki, H., eds. (1983): Prediction theory and harmonic analysis. North-Holland, Amsterdam.

Mason, J., Treas, R.S. (1986): Numerical weather prediction. Proc. Roy. Soc. A **407**(1832), 51–63.

Takens, F. (1981): Detecting strange attractor in turbulence. Lecture Notes in Math. **898**, Springer-Verlag, Berlin, 366.

Teodorescu, D. (1989): Time series decomposition and forecasting. Int. J. Control **50**(5), 1577–1596.

Wiener, N. (1950): The extrapolation, interpolation and smoothing of stationary time series. N.Y.

Wiener, N., Masani, P. (1957): The prediction theory of multivariate stochastic processes. Acta Math. **98**, 111–150; (1958): **99**, 93–137.

Parsimony in Dynamical Modeling

Alistair I. Mees and Kevin Judd

Abstract

Reconstruction of a dynamical system from data is in one sense easy: if we have enough parameters then we can exactly reproduce the original observations, and what could be better? The answer is, of course, that a lot of things could be better. We are usually interested in cases which are not quite the same as the observations, and if we try too hard to fit the data then we are probably fitting noise and making it likely that our predictions will be bad – maybe very bad. Even without noise, if our model is based on inaccurate assumptions, the results can be poor. In this chapter we discuss what it means to build a model, and describe some experience with a particular approach, i.e., the application of minimum description length to radial basis neural networks.

1 Dynamical Reconstruction

Given measurements from a dynamical system, it is natural to ask questions such as "Is this system chaotic?" or even "Can we discover the bifurcations or other underlying mechanisms that give rise to the dynamics?" In practice, there are always external influences or un-modelable internal effects and it is not always clear what these questions even mean. The embedding theorem [Packard et al., 1980; Takens, 1981; Noakes, 1991] makes it possible under some circumstances to *observe* the system state in the sense of control theory [Jacobs, 1974] and hence, using functional approximation methods such as radial basis functions [Casdagli, 1989] or tesselations [Mees, 1990; 1991], to "reconstruct" dynamics. As interpreted in this contribution, the reconstruction process builds a discrete dynamical system – a map on a Euclidean embedding space – that has approximately the same dynamics as the data.

This is only part of an ambitious research program which aims, among other things, to understand the effect of modeling and computational errors on nonlinear dynamics and to be able to distinguish chaos from noise, or to say why this cannot be done. There are several criticisms of this general program, in addition to any criticisms of the detailed methodology of any one approach. For example, the models are purely phenomenological, relying only on the data and not on any other knowledge about the system. How can we incorporate other things we might know about the system into our models? Once a model has been built, what are we to do with it? On the surface, the model appears to be more like an experimental system than a conventional mathematical model, in that it can be used for black-box simulation but perhaps not for much else. And finally, most approaches do not pay attention to known results about parsimony: they often use too many parameters, and tend to fit the noise as well as the dynamics, reducing their ability to predict and to generalize to other data sets from the same process.

Here we set out to show that at least some of these criticisms can be overcome: It is possible to make small (parsimonious) models, and there is even a good criterion for choosing the model size. The defect that the models are only black boxes can also be overcome [Glover and Mees, 1992; Judd and Mees, 1995], though we only touch on that here.

We follow the approach of Judd and Mees [Judd and Mees, 1994] in constructing radial basis function models, which are special types of neural networks with nice properties such as ease of use and assurance of global optimality once certain parameters (mainly the locations of the so-called *centers*) have been fixed. We do not go into detail on the modeling method, nor do we claim that the modeling method is necessarily the best, only that it is adequate for our purposes here and that it seems to be powerful in many cases. Our main concern is to make sure that we are unlikely to be fitting noise instead of deterministic dynamics, but the approach we take also helps answer other interesting questions, such as whether this model is any better than a linear model.

We begin by reviewing the radial basis method of functional approximation in the context of a more general class of *pseudo-linear* models. Then we discuss some general issues of nonlinear dynamical modeling, concentrating on how to build small models, and how to choose the model size. At the end we give some examples.

1.1 Radial Basis Function Modeling

How should we approximate a nonlinear function ρ from high dimensional space to the reals? In practice, "high" means 2 or more dimensional; nonlinear modeling in any dimension beyond 1 is difficult. A good starting point is to use *pseudo-linear models*, which are linear combinations of nonlinear functions. A pseudo-linear approximation for a map $\rho : \mathbf{R}^p \to \mathbf{R}$ is $\hat{\rho}$ where

$$\hat{\rho}(z) = \sum_{s=1}^{N} \lambda_s \phi_s(z) \tag{1}$$

for some arbitrary set of functions ϕ_s. Many approximation methods fall into this general class, but in this paper we will concentrate on radial basis (RB) functions, although there are cases where, for example, geometrical information can be got more readily from triangulation and tessellation methods [Watson, 1981; Mees, 1990, 1991].

One attraction of a pseudo-linear approximation is that its models can be manipulated just like conventional analytic models [Glover and Mees, 1992], enabling them to be analyzed using conventional analytic tools as well as computer simulations. The point is that exact derivatives of any order are available for the model, even though calculation of their numerical values requires use of the parameters stored in the computer model. Actually, the derivatives themselves can be calculated efficiently and accurately in the program without use of a separate symbolic manipulation package, by the techniques of "automatic differentiation", [Griewank and Corliss, 1991].

The radial basis function approximation for a map $\rho : R^p \to R$ is $\hat{\rho}$ where

$$\hat{\rho}(z) = \alpha.z + \beta + \sum_{s=p+2}^{N} \lambda_s \phi(|z - c_s|/r_s) \tag{2}$$

for a given *basis function* ϕ and *centers* $\{c_s\}$ and *radii* $\{r_s\}$, $s = p+2, \ldots, N$. Two of many possible basis functions are the cubic function $\phi(r) = r^3$ and the Gaussian function $\phi(r) = \exp(-r^2/2)$; there are many others, but a discussion of the best choice of functions would lead us too far afield. One of these functions is increasing and the other decreasing, but increasing and decreasing functions work about equally well. A useful image is to think of an RB model built with increasing basis functions as an elastic sheet which is fixed in place at a number of centers and which has some designated behavior at infinity; the more rapidly $\phi(r)$ increases with r, the stiffer is the sheet. A model built with decreasing functions can be thought of as a superposition of Gaussian humps. For a more complete discussion, see [Powell, 1985; Smith, 1992; Mees, Jackson and Chua, 1992]. The radial basis model (2) when viewed as a pseudo-linear model includes a constant function β, projections onto the components of z, which are all linear functions, and the nonlinear functions $\phi_s(z) = \phi(|z - c_s|/r_s)$. It is convenient to treat the constant and linear functions as particular choices of pseudo-linear function[12] and to use the general pseudo-linear form (1) from now on.

Presumably the reason we are approximating the map ρ is that we don't know ρ, but only some data $\{y_t, z_t\}$, $t = 1, \ldots, T$ where

$$y_t = \rho(z_t) + \nu_t$$

and ν_t is some noise or error term. To fit the model, we must solve

$$y_t = \sum_{s=1}^{N} \lambda_s \phi_s(z_t)$$

for all t, which is the same as solving the linear equations

---

[12]In fact, $\phi_i(z)$ for $i = 1, \ldots, p$ is projection onto the ith component of z and $\phi_{p+1}(z) = 1$ for all z.

$$\boldsymbol{Y} = \boldsymbol{\Phi}\lambda \tag{3}$$

where $\boldsymbol{Y}$ is the vector with elements y_t, $\boldsymbol{\Phi}$ is the T by N matrix with elements $\Phi_{ts} = \phi_s(z_t)$, and λ is the vector with elements λ_s. Usually $T > N$ and we solve (3) by least squares [Broomhead and Lowe, 1988, Chen, Cowan and Grant, 1991, Smith, 1992, Mees, 1993].

To get started, we have to find some way of choosing the centers $\{c_s\}$ and radii $\{r_s\}$. Most workers seem to choose $r_s = \sigma$ for all s and some σ calculated from the overall scale of the data, and choose centers randomly from among the data. This works quite well and is always a good starting point. Elsewhere [Mees, 1994a; Judd and Mees, 1994] the present authors have described powerful methods for doing the job better. Methods vary between two extremes: the quick-and-dirty approach that generates a large number of basis functions ϕ_s, more or less randomly, and uses the algorithm described in Sect. 1.3 to select the best basis functions to use; and at the other extreme sophisticated "reflux" algorithms that, given a set of basis functions, will generate a large number of new random basis functions that are likely to lead to better models, then apply the selection algorithm to the whole set and iterating until convergence is achieved. The sophisticated algorithm is only needed for complex problems.

An example of a problem where the quick-and-dirty algorithm is adequate is the following. Figure 1(a) shows the graph of a map $\rho : \boldsymbol{R}^2 \to \boldsymbol{R}$, the x coordinate of the Ikeda map [Casdagli, 1989]. Figure 1(b) shows an approximation $\hat{\rho}$ generated from a small radial basis model, with random choice of centers. The data in this case were points on a grid in $\boldsymbol{R}^2$ and their images under the map. It is apparent that the approximation is good for such a small model.

Most of the examples we give in this paper use a quick-and-dirty algorithm but Fig. 2 illustrates results of applying a sophisticated algorithm to a difficult problem. In this example the function ρ modeled is the luminance of an image and a radial basis model with ϕ Gaussian basis functions was used. Figures 2(a) and (b) show the original and the fitted image, while Fig. 2(c) shows the positions and radii of the centers chosen for the model.

1.2 Embeddings and Reconstructions

Suppose we are given a single-channel time series $\{y_t\}$, $t = 1, \ldots, \tilde{T}$ where $y_t \in \boldsymbol{R}$. An example is shown in Fig. 3. The time series is the x coordinate of an orbit of the Ikeda map [Casdagli, 1989] with added observational noise.

We assume that the observations come from a discrete-time dynamical system, with invisible state x_t and observed value y_t such that

$$x_{t+1} = f(x_t) \tag{4}$$

and

a

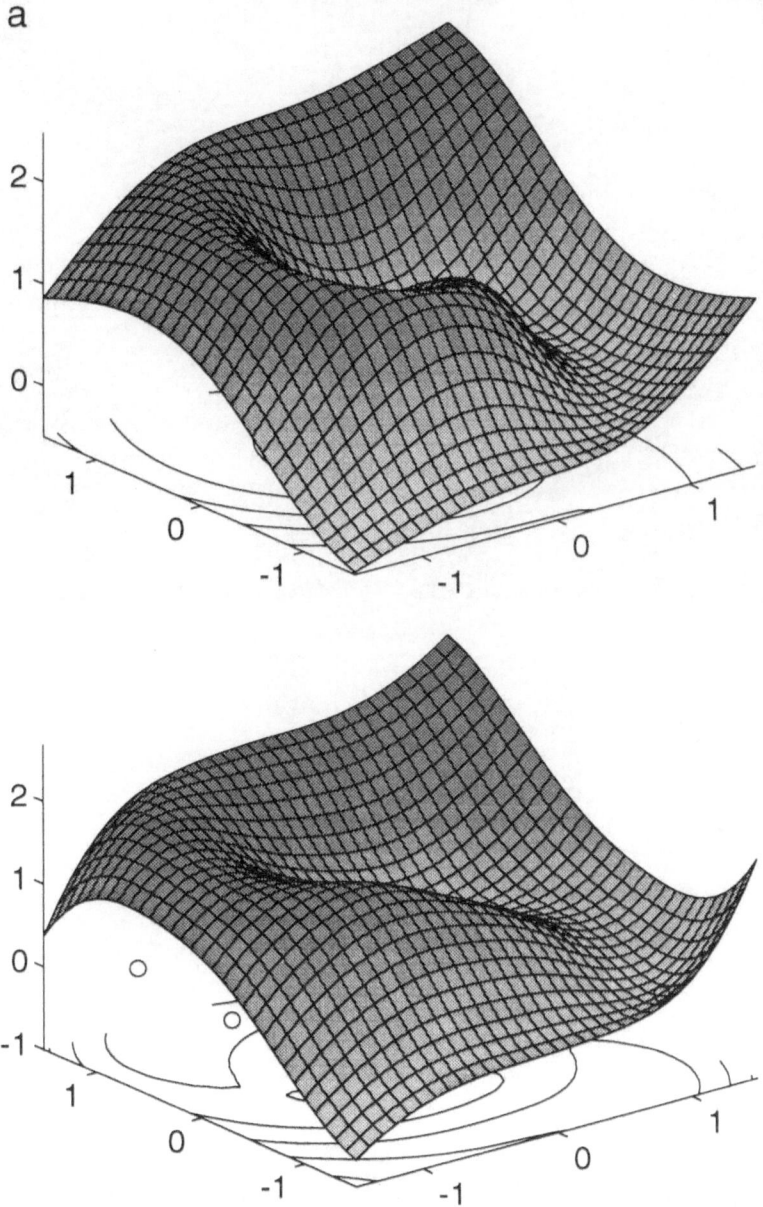

Fig. 1 (a) The x coordinate of the Ikeda map, which is a nonlinear trigonometric map on $\mathbf{R}^2$, shown as a surface plot overlaid on a contour plot. (b) A radial basis approximation to the surface in (a) based on 10 randomly chosen centers. Some of the centers are visible as small circles underneath the surface.

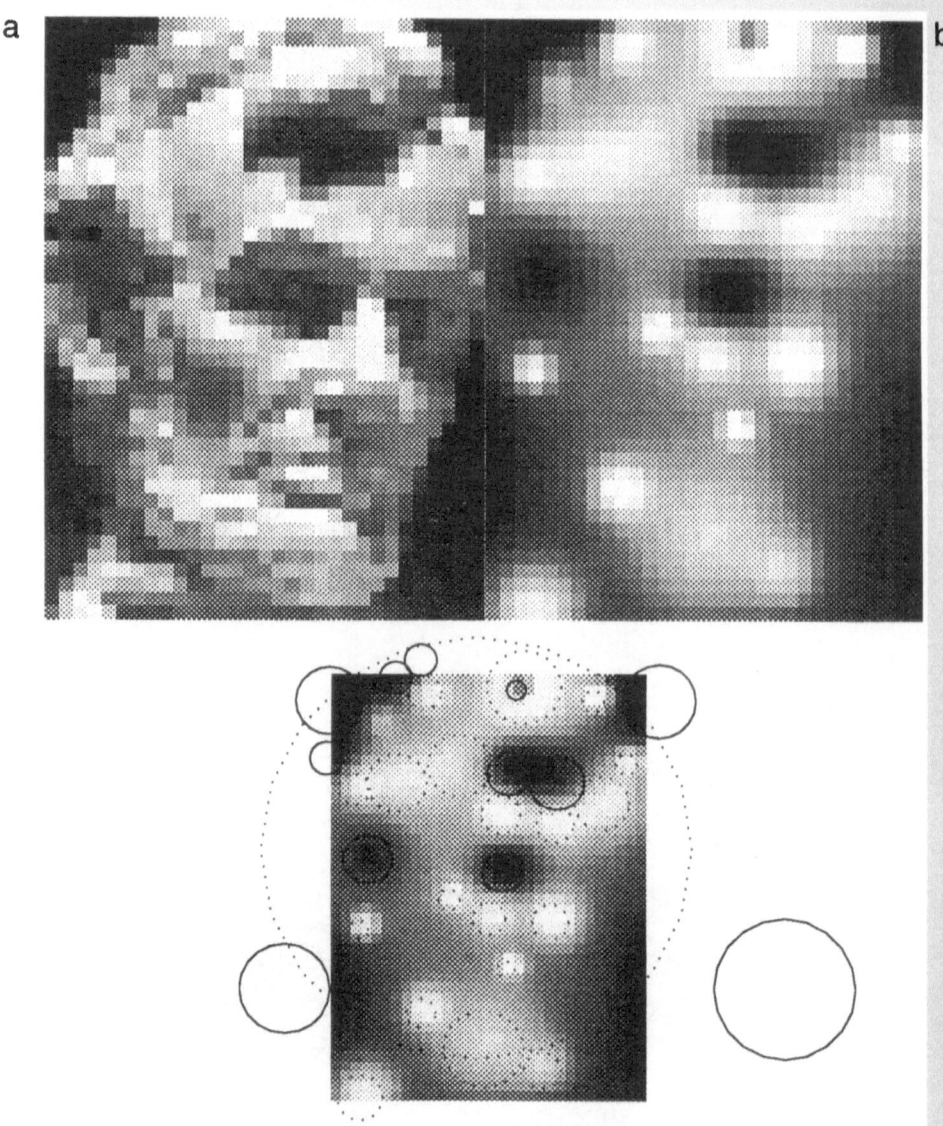

Fig. 2 A well-known American. (**a**) The original data. (**b**) The image fitted by a radial basis function. Unfocusing one's eyes and standing back helps in seeing the likeness. The fit does not look good to our eyes because the model minimises the sum of squares error, but our eye uses a different measure of closeness that puts more emphasis on edges. (**c**) The positions and radii of the centers used. Solid circles correspond to positive λ_s and dotted circles to negative λ_s. Notice how the optimal model chooses circles of various radii, some quite far from the image.

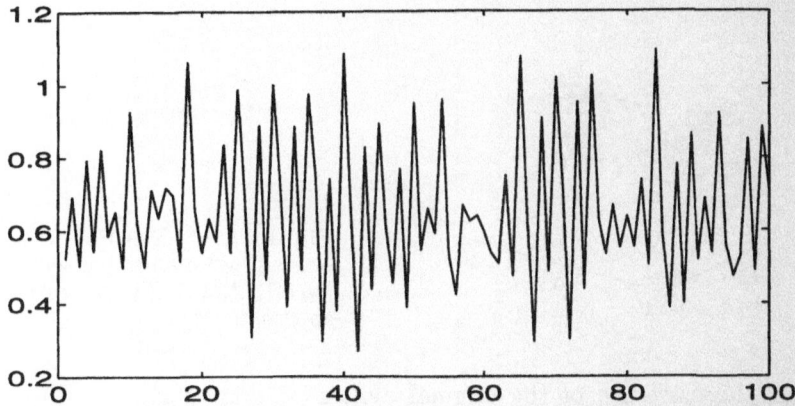

Fig. 3 Part of a noise-corrupted time series from a nonlinear dynamical system. The system is Ikeda's map; the time series is the x coordinate with observational $N(0, 0.05)$ noise added.

$$y_t = c(x_t) + \nu_t, \tag{5}$$

where ν_t are disturbances whose dynamics, if any, are unknown. We have made the assumption that the noise only affects the observations, not the dynamics. If there were a noise term in (4) we would have a harder problem with dynamical noise, though a more realistic one in many cases. This over-simplification may be remedied in part by work currently in progress [Mees and Smith, 1995].

We begin by choosing an embedding, such as the usual time-delay embedding given by

$$z_t = (y_t, y_{t-\tau}, \ldots, y_{t-(p-1)\tau}) \tag{6}$$

for some embedding dimension p and lag τ. The subject of embeddings has been discussed elsewhere in detail [Packard et al., 1980; Takens, 1981; Noakes, 1991; Sauer, Yorke, and Casdagli, 1992] and we do not go into it further here. For advice on choice of embedding dimension and related matters see [Albano et al., 1990; Broomhead, Jones, and King, 1986; Cheng and Tong, 1991]. Here, we simply assume that we have derived from the given 1-dimensional time series $\{y_t \in \mathbf{R}\}$, $t = 1, \ldots, \tilde{T}$ a p-dimensional time series $\{z_t \in \mathbf{R}^p\}$, $t = 1, \ldots, T$. (To simplify the notation, we are re-labeling the indices in $\{z_t\}$ so that they run from 1; if the embedding is as in (6) then $T = \tilde{T} - (p-1)\tau$.)

Figure 4 shows a 3-dimensional embedding of the noisy Ikeda data of Fig. 3. The dimension was selected by the method of false nearest neighbors [Abarbanel and Kennel, 1992].

Takens's theorem [Takens, 1981; Noakes, 1991] and its generalizations [Sauer, Yorke, and Casdagli, 1992] can be used to show that under certain

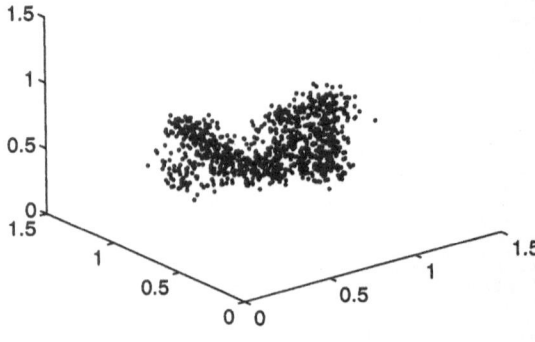

Fig. 4 The data of Fig. 3 embedded in 3 dimensions with a lag of 1. Not much structure is apparent.

conditions, the existence of the original dynamical map f implies [Mees, 1994b] there is a smooth dynamical map F on the embedding space:

$$z_{t+1} = F(z_t).$$

If we can provide an estimate $\hat{F}$ of F we have a model of the dynamics of the system and we can predict, interpolate, regulate, and in general apply the scientific method to any function of the observable which can be derived from the embedding. In particular, if the model is a radial basis model, we can, as mentioned earlier, find derivatives and locate fixed points of various periods, determine their stability and so on. The question of whether these fixed points, or any other attractors in the reconstructed system, correspond to equivalents in the "true" system is a subject of current research.

Usually it is not necessary to find $\hat{F}$, since we are only interested in the behavior of the observable y, rather than the reconstructed states z. It is sufficient to estimate the map $\rho : \mathbf{R}^p \to \mathbf{R}$ such that

$$y_{t+1} = \rho(z_t).$$

In the simple time delay embedding, ρ would be the first component of F. Estimating ρ is no harder than estimating F, and is usually easier.

Now we make a simplifying assumption. We pretend that the time series embedding coordinates z_t are known exactly, although the observed variables y_t may be subject to disturbances. This is incorrect since the embedding coordinates are derived from the observables, so that any disturbances on $\{y_t\}$ affect $\{z_t\}$, but for several reasons, including the fact that the embedding theorem is usually proved in the absence of noise, this assumption is implicit in virtually all nonlinear modeling. This problem will be returned to in a later paper [Mees and Smith, 1995]. The reason for neglecting it at present is that, although the far easier linear modeling problem can be set up [Kendall and Stuart, 1961] to account for noise in the domain as well as the range of the function, the presence of nonlinear terms makes calculation of covariances more difficult and may complicate the straightforward but powerful algorithms of Sect. 1.3. In fact, it is known that if the noise amplitude is small compared with the range of values of z_t, the assumption we are making here is relatively harmless [Draper and Smith, 1981]. We shall see later in this paper that in at least one example, the models are good even with the simplified noise model.

We therefore have the following problem: given a real-valued time series $\{y_t\}$ and a p-vector valued time series $\{z_t\}$ where it is assumed that

$$y_t = \rho(z_t) + \nu_t, \tag{7}$$

find an estimate $\hat{\rho}$ of ρ. This is exactly the problem treated in Sect. 1.1; since the approximation there minimized the mean square error, it is appropriate if the distribution of ν_t is normal and has zero mean. Throughout this contribution we shall assume that the disturbances are independently identically normally distributed random variables with zero mean and unknown variance. If they have nonzero mean, we absorb the mean into $\hat{\rho}$.

1.3 Models from Subsets of Centers

The representation in Sect. 1.2 allows us to apply pseudo-linear modeling methods, such as that in Sect. 1.1, for any fixed size of model and selection of centers and scales. Suppose we have a large candidate set of centers and scales (and possibly basis functions too). Even if we have agreed on a model size, which centers should we choose? Is there a way to decide which affine terms (if any) should be included?

We show in [Judd and Mees, 1994] how to apply ideas from nonlinear programming to this subset selection problem. The idea is to apply sensitivity analysis to a quadratic program derived from the fitting problem, and select good candidates to enter or leave the current subset. The insertion or removal is done by a method akin to the pivoting operation of the simplex algorithm of linear programming [Luenberger, 1984].

Given any subset I of indices, we successively add and remove indices from I until there is no further improvement in the fit of the model to the data without increasing the model size beyond its current value m. We use the following algorithm. In the description, $\hat{\Phi}$ is the matrix Φ of equation (3) with its columns normalized to unit length, the notation $\hat{\Phi}^I$ means the sub-matrix of $\hat{\Phi}$ with columns labeled by I and λ_I means the vector of corresponding elements λ_i, $i \in I$.

1 Start with any set I of size m.
2 For the current I, find λ_I as the least squares solution of $\hat{\Phi}^I \lambda_I = Y$.
3 Find $\epsilon = Y - \hat{\Phi}^I \lambda_I$ and $\mu = -\hat{\Phi}^T \epsilon$.
4 Find the index q that maximizes $|\mu_q|$.
5 Bring q into I and recalculate λ_I.
6 Let q' be the index which minimizes $|\lambda_i|$. Remove q' from I.
7 If $q' = q$, or the current subset I has previously been selected, terminate. Otherwise, return to step 1.

In practice, we start with a model of size $m = 0$ and enlarge it to size $m = 1$ by using steps 2–4 of the algorithm. Applying the algorithm is unnecessary at this stage: we already have the optimal size 1 model. Now enlarge from $m = 1$ to $m = 2$ using steps 2–4, and then use the algorithm to select a model of good size 2 starting from the one we have just generated. Continue in this way until we have a large enough model or a good enough fit or we have some other indication to stop such as the method discussed in Sect. 2.

2 Model Size and Type

We now have a method of representing time series data as a nonlinear fitting problem and a powerful fitting method, but questions remain, most notably that of what model we should be fitting to in the pseudo-linear algorithm of Sect. 1.1. One way to answer this question would be to try a number of models, with gradually increasing numbers of parameters, and somehow decide which is best. This is the approach taken here: we choose a model of size m from a finite but large class consisting of all choices of m centers from a relatively large number of candidate centers, and use the ideas outlined earlier to choose a subset of that size which gives a satisfactory model. The size m is gradually increased until it is large enough according to some criterion.

So let us assume we know how to make a model of given size, but not what size to choose. To choose a good size, we are going to use Rissanen's concept of minimum description length [Rissanen, 1989]. Let us outline the philosophy behind this briefly. Influenced by the work of [Chaitin, 1987], Rissanen considers a model to be a way of compressing the given data (say, for storage or for transmission over a computer network). The compression has to be lossless: that is, the data can be reproduced exactly from the compressed version. One reason for doing so is that one can argue that there is no such thing as a "correct" model; for example, one might obtain the best compression available by current technology, then notice later that the compressed data is every 13th digit of π in a certain base; the preceding phrase now becomes a very compact model which may even lead to new insights. By never throwing away information, we make it possible for someone to come along later and improve our work.

Consider the modeling process as a compression method. First we transmit information about which model we are going to use, from classes previously agreed between transmitter and receiver. Then we transmit the model's parameters, which may be variable in number. Finally, because the compression is to be lossless, we transmit the modeling errors in such a way that the receiver can reconstruct the actual data with complete accuracy.

The model must explicitly or implicitly have assumed a distribution for the errors: least squares fitting implies we are assuming Gaussian errors. Because the errors have an assumed distribution, they can be encoded efficiently: for example, if the model claims they are independent and defines the distribution they come from, it can use a Huffman code [Goldie and Pinch, 1991].

Using the methods outlined above, building a model of a given class and size can be done optimally or near-optimally. Similarly, the Huffman code is optimal. Thus the resulting total code length in bits, say, is in some sense optimal. It is called the *description length* of the data under the model.

The *best* model among all model sizes and all members of the agreed classes is then defined to be the one with minimal value of the description length. The idea of using minimum description length (MDL) to select models seems to be useful in many areas, such as in cluster analysis, in selection of numbers of bins for histograms, in regression and elsewhere [Rissanen, 1989].

The description length for any given model with m variables can be calculated directly from the definition [Judd and Mees, 1994] as follows. For the first part of the two part code, the model transmission, we have to describe the model $\hat{F}$, so we must first transmit a program for calculating it. For simplicity, we assume here that all programs are the same length; this is equivalent to transmitter and receiver agreeing beforehand on a fixed number of model classes, assumed equally likely to be used, so that the transmitter need only send the label of the class actually selected and the values of any parameters, say $\lambda_1 \ldots, \lambda_m$. This will also allow the receiver to construct the error distribution claimed by the model and so to decode the Huffman-encoded data that follows in the second part of the two-part code. The parameters are real numbers, so we will have to truncate them in order to transmit them in a finite code length. The key to the use of the MDL method in our problem is to realize that we can optimize over the truncation.

Rissanen's approach assumes a prior probability distribution on the parameters. (He argues that the final results are not at all sensitive to the choice of prior.) In the absence of other information, we choose the prior corresponding to the computer representation of floating point numbers: roughly, this corresponds to an exponential distribution centered around 0. We choose to send the jth parameter λ_j to a certain relative accuracy δ_j, so we actually send $\bar{\lambda}_j$ which contains only the first $\log_2 \delta_j$ bits in the fractional part of the normalization of λ_j. We show elsewhere [Judd and Mees, 1994] that the code length needed to specify the parameters $\bar{\lambda}_j$, $j = 1, \ldots, m$ is

$$L(\bar{\lambda}) \approx \sum_{j=1}^{m} \log \frac{\gamma}{\delta_j}$$

where L is in bits if logarithms are in base 2. The constant γ is not critical and represents the number of factors of 2 required in the exponent of a floating-point representation of a parameter: $\gamma = 32$ is more than adequate for nearly all purposes, and smaller values can be chosen if desired. Since it is common to work with natural logarithms in estimation theory, we shall do so from now on, and the code length will therefore be in "nats" rather than bits.

The total description length for the model and errors for a given set of data $\{z_t = (y_t, x_t)\}$, $t = 1, \ldots, T$, is

$$L(z, \bar{\lambda}) = L(z|\bar{\lambda}) + L(\bar{\lambda}) \tag{8}$$

where the data code length

$$L(z|\bar{\lambda}) = -\ln P(z|\bar{\lambda})$$

is just the negative log-likelihood of the data under the assumed distribution and the truncated parameter values. The MDL principle requires us to minimize (8) over $\bar{\lambda}$.

Building on the work of [Rissanen, 1989] we show elsewhere [Judd and Mees, 1994] that, to a good approximation, the optimal parameters are the maximum likelihood values truncated according to the (unique) solution $\hat{\delta}$ of

$$(Q\delta)_j = 1/\delta_j \tag{9}$$

where $Q = D_{\lambda\lambda}L(z|\hat{\lambda})$ is the second derivative matrix corresponding to the maximum likelihood solution. Substituting this into the total description length gives rise to the following formula for the description length of a model of size m:

$$S_m(z) = \left(\frac{T}{2} - 1\right) \ln \frac{\hat{e}^\top \hat{e}}{T} + (m+1)\left(\frac{1}{2} + \ln \gamma\right) - \sum_{j=1}^{m} \ln \hat{\delta}_j + C \tag{10}$$

where C is independent of the parameters and in fact is given by

$$C = \frac{T}{2}(1 + \ln 2\pi) + \frac{1}{2} \ln \frac{T}{2}.$$

Thus a good approximation to $\bar{\lambda}$ is the maximum likelihood value $\hat{\lambda}$, truncated according to the precision specified by the solution of (9), resulting in a description length bounded by $S_m(z)$ given by (10). We shall minimize $S_m(z)$ to choose a good model. The only difference between minimizing this approximate version of MDL and using the conventional maximum likelihood method is that we have to account for the additional penalty terms defining the truncation. But it is precisely these terms which enable us to choose the appropriate size of model, since, if we increase the number m of nonzero parameters, the sum of squares in (10) always decreases but the cost of the parameters generally increases. (Since in this introductory presentation we are not distinguishing carefully between the number m of parameters and the number of bits in the model, it is always possible that a model with more parameters could be more compactly represented even without the errors, if the parameters did not need to be specified to great precision.)

The precision $\hat{\delta}_j$ has the useful property of being the minimum required precision of λ_j. This specification of parameters is perhaps more useful than

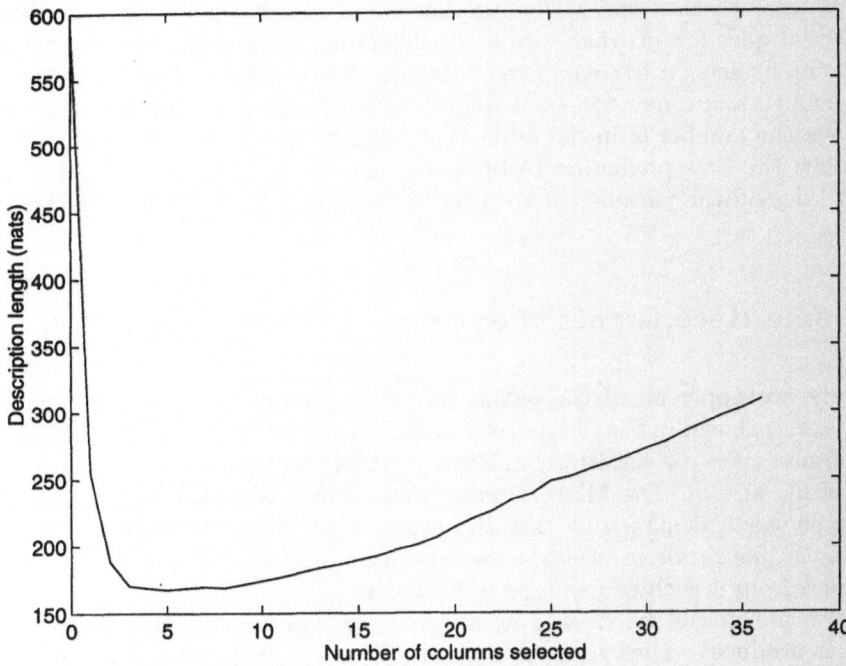

Fig. 5 A plot of description length against size of model using the annual average sunspot number time-series embedded in 3 dimensions with a lag of 1. The model was build from an initial set of constant and linear functions and 1000 Gaussian radial basis functions. The optimal model used the current year's sunspot number and 4 radial basis functions.

stating standard deviations, because it gives an unequivocal range over which the parameters can vary and the model remain accurate. The precision may be also be used to indicate that some parameter is actually not useful, if its precision is comparable to its value. (If we were fitting models with global optimality this would not be necessary, but our fitting process may not always be accurate.)

It is clear that the MDL criterion is related to other well-known model selection criteria, and Rissanen shows that asymptotically, the above approximate expression for MDL is equivalent to the Schwarz (or Bayesian) information criterion [Tong, 1990; LeBaron, 1991; Mees, 1993; Schwarz, 1978]. We have found, however, that working with the above form gives better results for smaller data sets, and for large ones in critical cases, and the extra computation required is insignificant.

An example is given in Fig. 5, which shows the variation in description length as the number of parameters is increased in a model of the sunspot series, discussed in [Judd and Mees, 1994].

We remark that the MDL criterion could also be used to answer the perennial question of what is a good embedding for a given reconstruction problem, by using it to compare models with different embedding dimensions. The embedding dimension (and lag) come in as parameters to be specified and via the number of initial terms that must be specified in the time series to allow the first prediction to be made, as well as in the specification of model-dependent parameters such as location of centers in a radial basis model.

3 Some Reconstructed Systems

Finally, we apply all of the results so far to an example. The embedded time series shown in Fig. 4 has 1000 points. A subset of size 50 was selected randomly; then the sensitivity analysis method was used to build models of increasing size m. The MDL criterion selected a model with $m = 18$. It is perhaps worth emphasizing that the model used was derived entirely from the noisy data, not from any knowledge whatever of the map or of other samples from it such as was used in Sect. 1.1.

The model can be treated as a dynamical system and allowed to run free; it produces a time series shown embedded in Fig. 6; notice the delicate structure in both parts of Fig. 6 which was not at all apparent in the noisy data of Fig. 4.

Actually, this picture is a little misleading: it turns out that it really shows a chaotic transient, and that eventually the model system's trajectory settles down to a period 8 orbit. In fact, with some input time series (having different noise realizations to the one used here) we get the period 8 orbit with shorter transients. This is scarcely surprising, given the sensitivity of chaotic systems to perturbations, and would still indicate that the noise-corrupted data originally came from a deterministic system which was in or near a chaotic regime, with an attractor in the correct part of the space.

The numerical ability of the model to predict one step turns out to be very good on average, with a correlation coefficient of more than 98%; see [Mees, 1994b] for details. Many more studies can be made on the model, such as looking at how the predictive ability varies with the number of time steps ahead to be predicted; this is related to the Lyapunov exponent of the chaotic attractor [Mees, 1991; Mees et al., 1992]. Space prohibits doing so here.

As a final example we illustrate how, by examining the bifurcations of a model, one can infer properties of an experimental system that are not directly observable because of measurement and dynamic noise. Elsewhere we have described how to infer a Shil'nikov mechanism as a chaos generator from a time-series [Judd and Mees, 1994]. Measurements from forced oscillations of a stretched string were modeled using the techniques described in this chapter; complete details appear elsewhere [Judd et al., 1995]. Figure 7

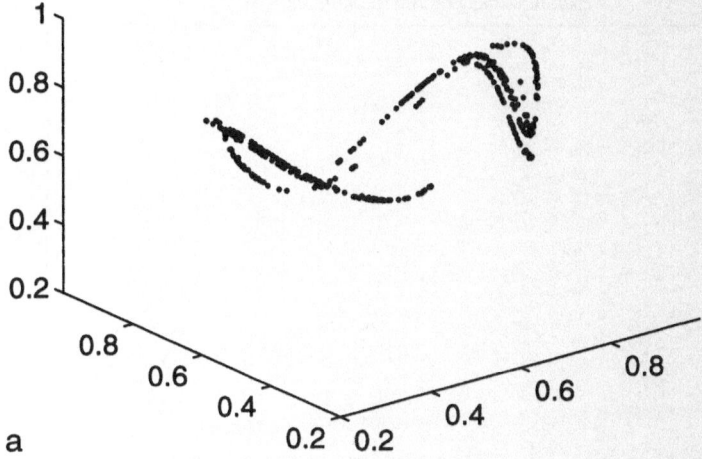

a

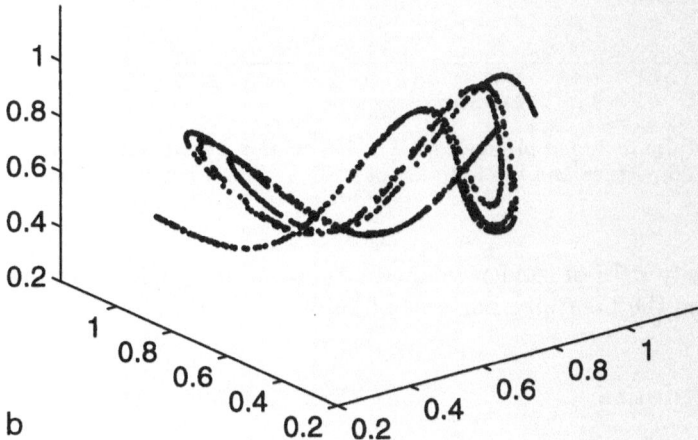

b

Fig. 6 (a) An embedding (3 dimensions, lag 1) of a trajectory from the model of the data of Fig. 4. The trajectory looks very like the real Ikeda attractor shown in (b) with the same embedding.

shows an experimentally determined "bifurcation" diagram. Figure 8 shows the bifurcation diagram of the reconstructed system, which clearly shows a period doubling cascade and period-three window not evident in Fig. 7.

There are many applications in progress or completed. The sunspot time series is discussed in [Mees, 1993] and [Judd and Mees, 1994], where the conclusion drawn is that the series is best modeled by a nonlinear periodic system with dynamic noise. A quite different application is an attempt, currently in

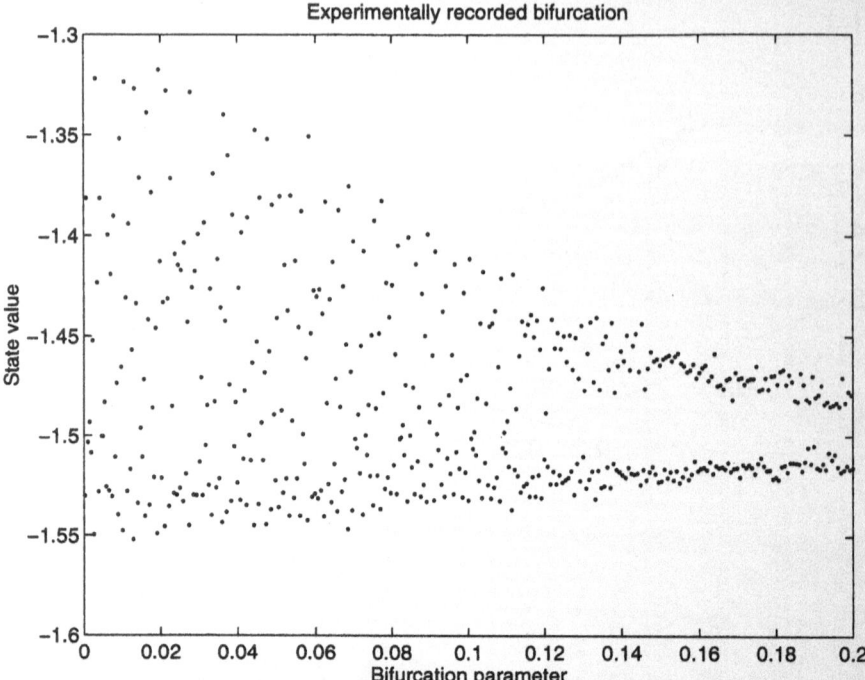

Fig. 7 Experimentally recorded bifurcation. The bifurcation parameter is a monotonic function of temperture and the function of state is a Poincaré section coordinate.

progress, to classify risks of sudden infant death syndrome [Judd and Stick, 1995] by modeling the breathing patterns of babies.

Acknowledgments

Financial support was provided by the Australian Research Council and the Faculty of Engineering and Mathematical Sciences, The University of Western Australia. The data for Fig. 2 were provided by Math Works, Inc., makers of the Matlab package which was used for all calculations in this contribution.

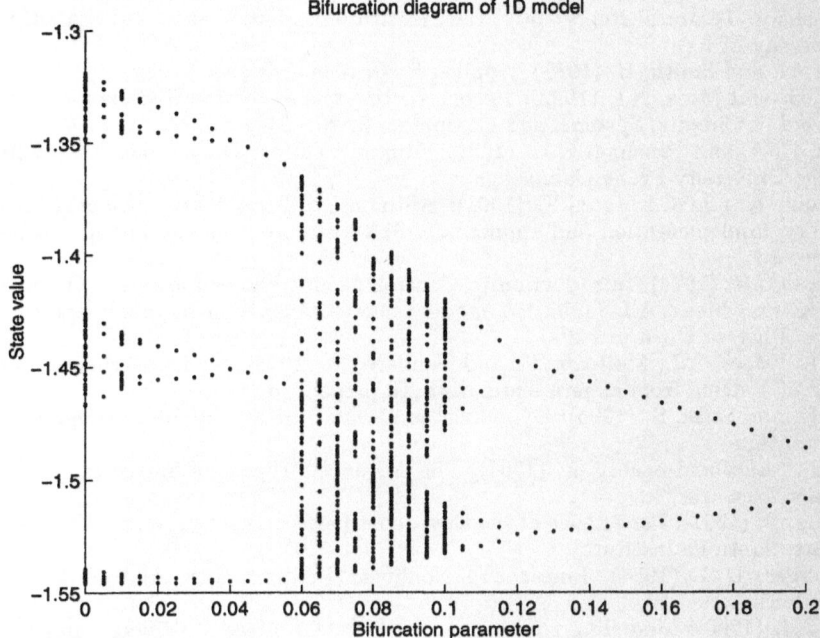

Fig. 8 Bifurcation diagram of one-dimensional model built from the experimental data. Clearly shown is a period doubling cascade and period-three window not evident in Fig. 7.

References

Abarbanel, H.D.I. and Kennel, M.B. (1992): Local false nearest neighbors and dynamical dimensions from observed chaotic data. Technical report, Department of Physics, University of California, San Diego.

Albano, A.M., Muench, J., Schwartz, C., Mees, A.I., and Rapp, P.E. (1988): Singular value decomposition and the Grassberger-Procaccia algorithm. Phys. Rev. A **38**, 3017–3026.

Broomhead, D.S., Jones, R., and King, G.P. (1986): Topological dimension and local coordinates from time series data. Technical report, Department of Mathematics, Imperial College, London.

Broomhead, D.S. and Lowe, D. (1988): Multivariable functional interpolation and adaptive networks. Complex Systems **2**, 321–355.

Casdagli, M. (1989): Nonlinear prediction of chaotic time series. Physica **35D**, 335–356.

Chaitin, G.J. (1987): Algorithmic Information Theory. Cambridge University Press, Cambridge.

Chen, S., Cowan, C.F.N., and Grant, P.M. (1991): Orthogonal least squares learning algorithm for radial basis function networks. IEEE Transactions on Neural Networks **2**, 302–309.

Cheng, B. and Tong, H. (1991): On consistent non-parametric order determination and chaos. Technical Report 6th draft, Institute of Mathematics and Statistics, University of Kent.

Draper, N. and Smith, H. (1981): Applied Regression Analysis. Wiley, New York.

Glover, J. and Mees, A.I. (1992): Reconstructing the dynamics of Chua's circuit. Journal of Circuits, Systems and Computers 3, 201–214.

Goldie, C.M. and Pinch, R.G.E. (1991): Communication Theory, Vol. 20. Cambridge University Press, Cambridge.

Griewank, A. and Corliss, G.F. (1991): Automatic differentiation of algorithms: Theory, implementation and application. SIAM Proceedings in Applied Mathematics 53.

Jacobs, O.L.R. (1974): Introduction to Control Theory. Clarendon Press, Oxford.

Judd, K. and Mees, A.I. (1994): A model selection algorithm for nonlinear time series. Physica D (in press).

Judd, K., Mees, A.I., Molteno, T., and Tufillaro, N. (1995): Modeling chaotic motions of a string from experimental data. In preparation.

Judd, K. and Stick, S. (1995): Dynamical models for infant breathing patterns. In preparation.

Kendall, M.G., and Stuart, A. (1961): The Advanced Theory of Statistics, Vol. 2. Hafner, New York.

LeBaron, B. (1991): Persistence of the Dow Jones index on rising volume. Technical report, Santa Fe Institute.

Luenberger, D.G. (1984): Linear and Nonlinear Programming. Addison-Wesley (London).

Mees, A.I. (1990): Modelling complex systems. In A.I. Mees T. Vincent and L.S. Jennings, eds., Dynamics of Complex Interconnected Biological Systems, Vol. 6 (pp 104–124). Birkhauser, Boston.

Mees, A.I. (1991): Dynamical systems and tesselations: Detecting determinism in data. International Journal of Bifurcation and Chaos 1, 777–794.

Mees, A.I. (1993): Parsimonious dynamical reconstruction. International Journal of Bifurcation and Chaos 3, 669–675.

Mees, A.I. (1994a): Nonlinear dynamical systems from data. In F.P. Kelly, ed., Probability, Statistics and Optimisation (pp 225–237). Wiley, Chichester, England.

Mees, A.I. (1994b): Reconstructing chaotic systems in the presence of noise. In M. Yamaguti, ed., Towards the Harnessing of Chaos (pp 305–321). Elsevier, Tokyo.

Mees, A.I., Aihara, K., Adachi, M., Judd, K., Ikeguchi, T., and Matsumoto, G. (1992): Deterministic prediction and chaos in squid axon response. Physics Letters A 169, 41–45.

Mees, A.I., Jackson, M.F., and Chua, L.O. (1992): Device modeling by radial basis functions. IEEE Trans CAS/FTA 39, 19–27.

Mees, A.I. and Smith, R.K. (1995): Estimation and reconstruction in noisy chaotic systems. In preparation, 1995.

Noakes L. (1991): The Takens embedding theorem. International Journal of Bifurcation and Chaos 1, 867–872.

Packard, N.H., Crutchfield, J.P., Farmer, J.D., and Shaw, R.S. (1980): Geometry from a time series. Phys. Rev. Lett 45, 712–716.

Powell, M.J.D. (1985): Radial basis functions for multivariable interpolation: a review. Technical Report 1985/NA12, Department of Applied Mathematics and Theoretical Physics, Cambridge University.

Rissanen, J. (1989): Stochastic Complexity in Statistical Inquiry, Vol. 15. World Scientific, Singapore.

Sauer, T., Yorke, J.A., and Casdagli, M. (1992): Embedology. J. Stat. Phys. **65**, 579–616.

Schwarz, G. (1978): Estimating the dimension of a model. Annals of Statistics **6**, 461–464.

Smith, L.A. (1992): Identification and prediction of low-dimensional dynamics. Physica **58**D, 50–76.

Takens, F. (1981): Detecting strange attractors in turbulence. In D.A. Rand and L.S. Young, eds., Dynamical Systems and Turbulence, Vol. 898 (pp 365–381). Springer, Berlin.

Tong, H. (1990): Non-linear Time Series: a Dynamical Systems Approach. Oxford University Press, Oxford.

Watson D.F. (1981): Computing the n-dimensional Delaunay tesselation with application to Voronoi polytopes. The Computer Journal **24**, 167–172.

The Bifurcation Paradox:
The Final State is Predictable
If the Transition Is Fast Enough

Oleg Ya. Butkovskii, Yurii A. Kravtsov and Jeffrey S. Brush

Abstract

The problem of symmetry breaking under bifurcation transition is investigated. It is shown that selection of one of two final states (which are usually assumed to be equiprobable) may be realized with a probability close to unity if the transition is carried out with a sufficient speed. The probability of the final state depends on both the speed of the transition and the noise level in the system. As an example, the transition through the first period-doubling bifurcation of the logistic map was investigated. It was found that the probability of transition to a given final state may vary from unity (corresponding to the fast noiseless transition with symmetry) to 1/2, when the transition is rather slow and results in a symmetrical distribution over possible states. The states corresponding to transition with symmetry breaking are shown to be separated from the equiprobable states by a rather narrow boundary which is determined by the relationship between the transition speed and the noise level. A number of physical situations are considered for the illustration of this bifurcation paradox.

1 Introduction

When nonlinear systems undergo bifurcation, they may end up in any of two (or several) new stable equilibrium states. Which of the states is observed during a particular trial has been typically considered a random variable, since the behavior of the system at the instant of bifurcation is determined by the influence of noise. In particular, if there are two possible new equilibrium states which only differ in phase (e.g. $s_1 = A\ B\ A\ B\ A\ B\ \ldots$ and $s_2 = B\ A\ B\ A\ B\ A\ \ldots$, for a period-two equilibrium state), the probabilities $P_1 \equiv P(f_1)$ and $P_2 \equiv P(f_2)$ of transition into each of them should be equal :

$$P(s_1) = P(s_2) = 0.5 \,. \tag{1}$$

However, in real systems the final state is determined not only by noise, but also by the speed of the transition through the bifurcation point. When system parameters are changed quickly, the noise process simply has no time to influence the resulting state of the system, and the transition occurs as if there were no noise at all. A noiseless bifurcation transition with finite speed turns out to be regular and fully predictable. In this case, the probabilities $P(f_1)$ and $P(f_2)$ of the transition into states f_1 and f_2 are solely dependent on the initial conditions and are either

$$P(f_1) = 0 \text{ and } P(f_2) = 1 , \quad \text{or} \quad P(f_1) = 1 \text{ and } P(f_2) = 0 . \tag{2}$$

The differences in behavior of a bifurcating system based on the absence or presence of dynamic noise motivated us to explore the regimes in which the probabilities given in (1) and (2) hold, and when some intermediate values prevail. We describe a relationship between the rate of change of a control parameter (the "speed" of transition through the bifurcation), the noise level, and the predictability of the final state.

For our analysis, we chose the relatively simple example of the first period-doubling bifurcation in the logistic map. The simplicity of the example will not detract from the general applicability of these ideas. We determine, for a given noise level, the transition speed needed to ensure that the post transition state of the system has probabilities similar to Eq. (2).

1 The Noisy Logistic Map with Time-Varying Control Parameter

We used as our system of study the well-known logistic map:

$$x(n+1) = Rx(n)[1 - x(n)] , \tag{3}$$

modified in two respects. First, we make the parameter R time varying, and second we introduce dynamic noise at each iteration. Thus, our system is:

$$x(n+1) = R(n)x(n)[1 - x(n)] + \eta(n, \gamma) , \tag{4}$$

where $\eta[n, A]$ is a simulated, zero-mean, pseudo-random noise process, uniformly distributed between $-\gamma$ and $+\gamma$. Its variance is $\sigma^2 \equiv \langle \eta^2 \rangle = \gamma^2/3$.

The first period-doubling bifurcation of the logistic system occurs near $R = 3.0$, and it is this bifurcation that we will study. Thus, we use for $R(n)$ the piecewise linear function:

$$R(n) = \begin{array}{lll} R_0 & \text{for} & n < N \\ R_0 + \Delta R(n - N_1)S & \text{for} & N_1 \leq n < N_2 \\ R_f & \text{for} & n > N_2 \end{array} \tag{5}$$

In Eq. (5), $R(n)$ is constant at R_0 until some time N_1, after which it steps up at each iteration step until N_2, after which it is constant at R_f. We used $R_0 = 2.8$ and $R_f = 3.2$, so $\Delta R = 0.4$. $1/S = N_2 - N_1$ is the number of time steps needed to complete the transition, so S is thus the "speed" of the system through the transition region (that is, the speed will be 1.0 when the

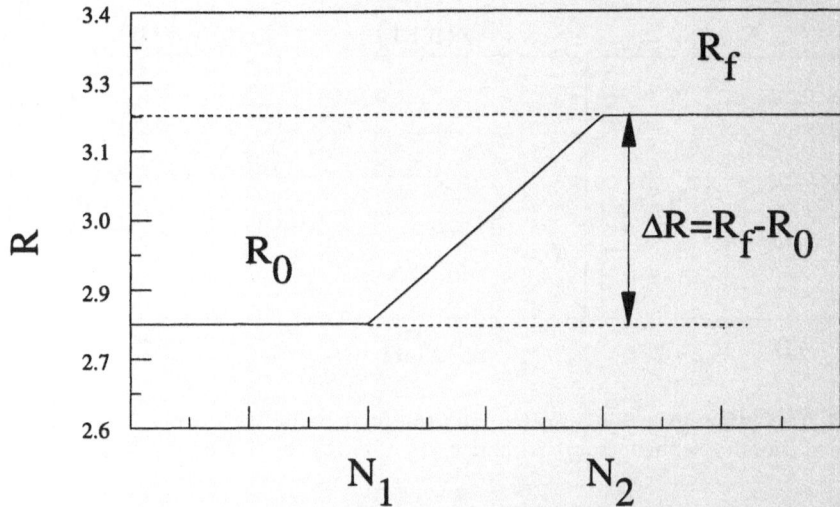

Fig. 1 An example of the dependence of the control parameter $R(n)$ on discrete time n.

transition occurs in one time step, 0.1 when the transition occurs in ten time steps, etc.). Figure 1 contains a typical example of $R(n)$.

In the absence of noise, once the parameter R has reached R_f, the system will be in a period-two regime in which $x(n)$ alternates between a high and a low value. In this regime, there are two possible final "states", which we denote f_1 and f_2. For f_1, $x(n)$ takes on the low value for odd indices and the high value for even indices, and vice versa for f_2. The states are thus related in that $x_{f_1}(n) = x_{f_2}(n+1)$. Figure 2 shows the logistic bifurcation diagram for parameters between 2.6 and 3.4, the limit cycle for $R = 3.2$, and representations of the two possible final states.

2 Transition Probabilities in the Presence of Noise

We analyzed the system described above using speeds S in the range from 0.0002 to 1.0 (transition times from 1 to 5000 steps), and values of γ between about 9×10^{-12} and 9×10^{-2}. Noise, uniformly distributed between $-\gamma$ and $+\gamma$, has a variance $\gamma^2/3$, and the logistic map with $R = 3.2$ has a variance of 0.0205. Taking the logistic system as our "signal", the signal to noise ratio (SNR, in dB) is then

$$\text{SNR} = 10\log_{10}\left(\frac{0.0205}{\gamma^2/3}\right) = -20\log_{10}\gamma - 12.1\,. \tag{6}$$

Therefore, our range of values for γ corresponds to SNRs between about 20 and 220 dB. For all trials, we used as an initial condition $x(0) = (R_0-1)/R_0 =$

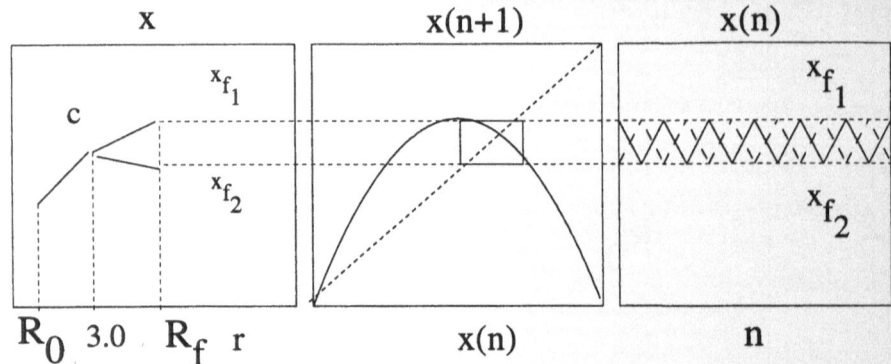

Fig. 2 The bifurcation diagram (a); (b) the final period-two orbit; and counter-phase stationary sequences $x_{f_1}(n)$ and $x_{f_2}(n+1)$ (c) for the logistic map

0.6429, which is the fixed point solution of Eq. (3), and generated additional iterates using the noisy logistic map, Eq. (4), and independent noise processes.

When S is unity (that is, when the parameter R is abruptly changed from R_0 to R_f in one time interval), the final state in the absence of noise is always f_1 (by definition). We denote the probability of observing f_1 as P_1, and our further analysis is based on the dependence of P_1 on noise and on S. P_1 was obtained by running 2500 bifurcation trials at each speed/noise combination and computing the fraction of those trials for which the final state was f_1. This fraction was taken as the estimate for P_1.

In Fig. 3, P_1 is plotted against noise level for two values of S (0.004 and 0.008) to demonstrate the typical response to increasing noise levels. Clearly, for very low noise (high SNR), P_1 is 1.0. As the noise level increases, P_1 eventually goes to 0.5 (that is, the probability of getting either final state is the same, and the transition is fully stochastic). We take as the critical noise level for a given speed the noise level for which P_1 is 0.75 (we call this critical noise level σ_{crit}). For noise levels below σ_{crit}, the system is mainly deterministic (P_1 is closer to unity than to 0.5) and for noise levels greater than this value the system is mainly stochastic (in the last case P_1 is closer to 0.5).

The critical noise level σ_{crit} is plotted against transition speed S in Fig. 4. A power law (linear on a log–log scale) relationship was fit to the data and is shown as well. The power law fit was:

$$\text{critical SNR (dB)} = 10.27 - 52.02 \log_{10} S, \tag{7a}$$

or

$$\log_{10}(S_{\text{crit}}) = 0.1974 - 0.01922 \, \text{SNR (dB)}. \tag{7b}$$

Using noise variance $\sigma_{\text{crit}}^2 = \langle \eta^2 \rangle = \gamma^2/3$ instead of SNR, Eq. (6), we may rewrite Eq. (7a) as

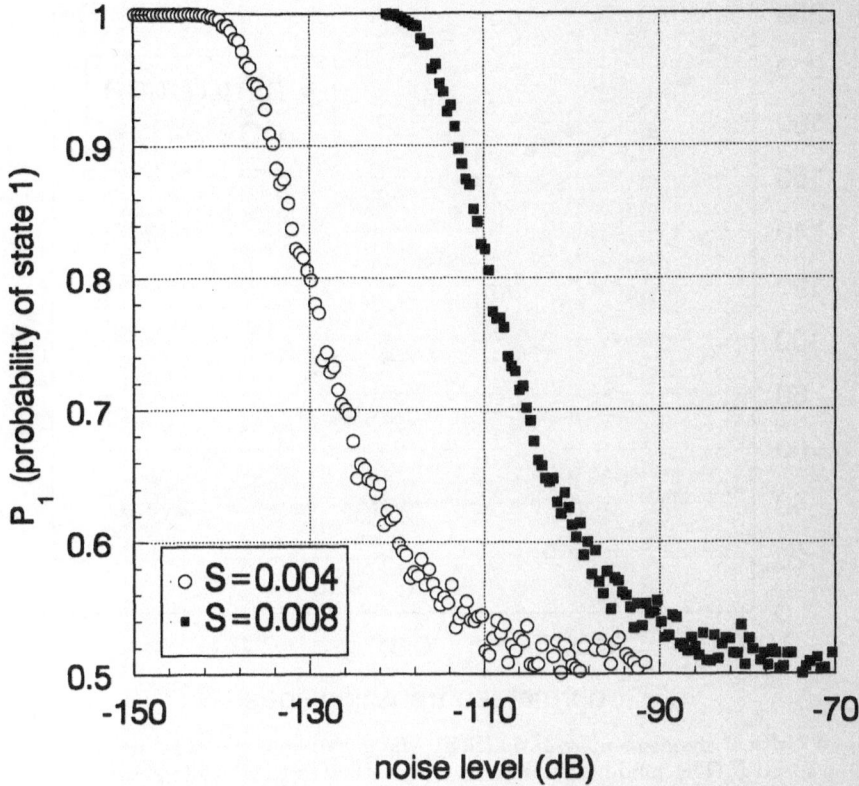

Fig. 3 P_1 vs. noise level σ^2 for two transition speeds, $S = 0.004$ (250 time steps) and $S = 0.008$ (125 time steps). Each symbol represents the fraction of trials (out of 2500) in which the final state was f_1. The sigmoid character of the curves is characteristic. For $S = 0.004$, the critical noise level is roughly 128 dB, while for $S = 0.008$ it is roughly 108 dB.

$$\sigma^2_{\text{crit}} = CS^a \qquad (8)$$

with $a \approx 5$.

The final state of a system undergoing a bifurcation is thus sensitive to the rate of transition as well as to the noise level. It is clear that for very low transition speeds, the critical noise level is also low, and rises with increasing speed S. When $S \gg S_{\text{crit}}$ for a given noise level, P_1 will be near unity, and the system can be considered deterministic, and when $S \ll S_{\text{crit}}$, P_1 will be near 0.5, and the system can be considered to be purely stochastic. The general recommendation for moving a noisy system through a bifurcation to a desired state is thus to do it as quickly as possible, rather than slowly and carefully!

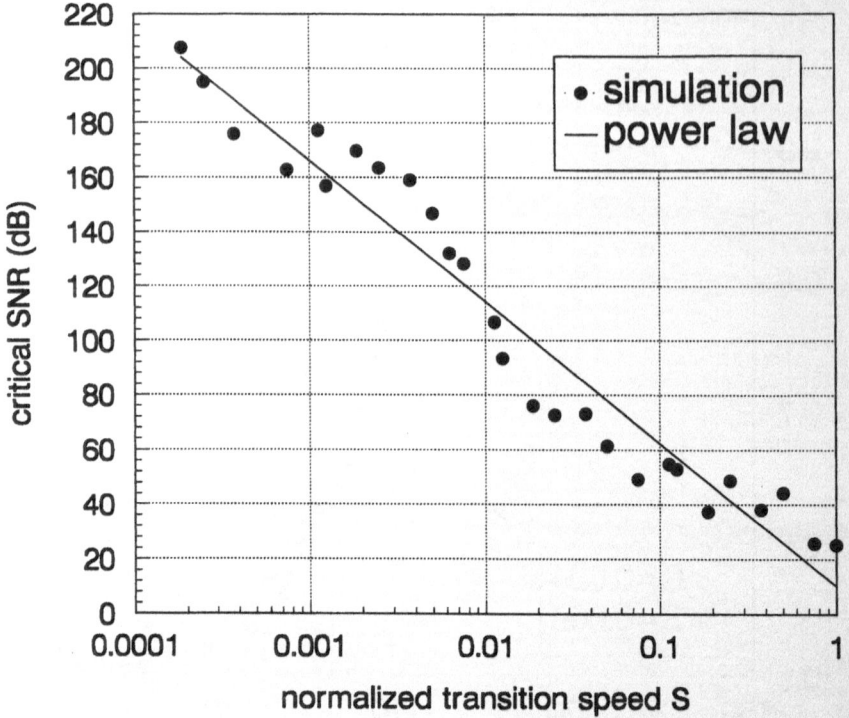

Fig. 4 Critical signal-to-noise ratio (SNR, the negative of the noise level) vs. transition speed S. The solid line represents a power law fit to the data shown (Eq. (7)).

3 The Bifurcation Paradox and Spontaneous Symmetry Breaking

The bifurcation paradox has a direct relationship to the problem of spontaneous symmetry breaking. When a slowly changing control parameter, say, the pump intensity in a laser system [Zheludev, 1989], exceeds the critical value, two stable polarization states arise with equal probability $P_1 = P_2 = 1/2$. In this adiabatic situation the final polarization state is random due to the influence of small fluctuations. The situation may drastically change if the transition is sufficiently rapid so that weak noise is not able to influence the final state. In this case the probability of a definite polarization state may be much higher than the probability of another state. At this stage of our knowledge we could hardly prove that breaking of polarization symmetry in a laser system is a result of high speed transition, but this hypothesis looks rather realistic.

In other systems with spontaneous symmetry breaking (spontaneous mirror symmetry breaking in nature, related to the origin of life [Goldanskii and Kuzmin, 1989]; symmetry breaking in population dynamics, in phase transition, and in morphogenesis [Horsthemke and Lefever, 1984; Nicolis and Prigogine, 1989]) the hypothesis of fast, non-adiabatic transition as a mechanism for violating the symmetry may also be promising as a new idea for future explanation.

4 Conclusion

The analysis carried out in this paper for the period-doubling bifurcation of the logistic map revealed some general regularities for the non-adiabatic transition of dynamical systems through the bifurcation point. They are as follows:

i) The probability for transition into the a given "post-bifurcation" state depends on the relation between the speed S of changing the control parameter and the noise level σ^2 in the system. With low noise level (or with rapidly changing parameter), when $\sigma^2 < \sigma^2_{\text{crit}} = CS^a$, the dynamic regime is realized, in which the probability of transition P_1 is near unity. In contrast, with relatively large noise level or slowly changing parameter, $\sigma^2_{\text{f}} > \sigma^2_{\text{crit}}$, the stochastic regime dominates, resulting in equiprobable states.

ii) For the noisy logistic map the index a in the power dependence $\sigma^2_{\text{crit}} = CS^a$ takes a rather high value $a \approx 5$. It is due to this fact that the dynamic regime of transition process is separated from the stochastic one by a rather narrow margin. Overall increasing the noise level makes the degree of predictability noticeably lower and shortens the time of predictable behavior.

In spite of the simplicity of the model chosen for the analysis, the relationships above are of rather general nature, probably generic with other values of the index variable than 5. The problem under consideration may apply to a great variety of bifurcation phenomena of physical, technological, or social nature. It is worth noticing that currently the process of quick economical reforming with minimal danger of producing economical bifurcation deadlock is of special interest for many countries. Our results might support the actions of resolute reformers who adhere to the principle "the faster the better", though every recommendation should be based on detailed calculations.

O. Butkovskii and Yu. Kravtsov are indebted to the International Science Foundation for the partial financial support of this work, under the auspices of NAG000 and NAG300.

References

Goldanskii, V.I., Kuzmin, V.V. (1989): "Spontaneous mirror symmetry breaking in nature and the origin of life". Sov. Phys. Uspekhi **32**(1).

Horsthemke, W. and Lefever, R. (1984): *Noise-Induced Transition. Theory and Applications in Physics*, Chemistry and Biology. Springer-Verlag, Berlin, Heidelberg.

Nicolis, G., Prigogine, I. (1989): *Exploring Complexity: An Introduction.* W.H. Freeman, New York.

Zheludev, N.I. (1989): "Polarization instability and multistability in nonlinear optics". Sov. Phys. Uspekhi **32**(4).

Part 4

Prediction of Biological Systems

Models and Predictability of Biological Systems

Martin P. Paulus

Abstract

The purpose of this paper is: firstly to emphasize relevant basic notions for models of prediction in biological systems; second, to sketch selected approaches that attempt to predict biological phenomena; third, to discuss the limits of, and problems with, predictability in biological systems; and last to provide two examples of approaches which attempt to determine the spatio-temporal scale for optimal prediction, as well as order parameters governing the predictability of sequences.

1 Prediction in Biological Systems

Prediction refers to the forecasting of an empirical event based on the observation of other events, via experience or scientific reasoning. In physical systems, the spatio-temporal evolution can be derived in many cases from first principles, or reduced to simple models that are governed by differential equations. Systems with many degrees of freedom are often embedded into a probabilistic framework, e.g. statistical mechanics. This framework provides methods for the reduction of the degrees of freedom to a few relevant parameters that predict the spatio-temporal evolution.

In contrast, the spatio-temporal evolution of biological systems appears to be neither completely reducible to deterministic equations derived from first principles nor completely probabilistic. Others [Haken, 1983] have long recognized the importance of both deterministic as well as stochastic processes for the production of macroscopic spatial, temporal, or functional structures. Moreover, distinct spatio-temporal patterns have been described in many biological systems ranging from low-dimensional attractors of neuronal firing [Freeman and Skarda, 1985; Doyon, 1992] to "chaotic" sequences in epidemics of infectious diseases [Schaffer, 1995].

Traditionally, predictive models in biology and medicine are based on a probabilistic framework that separates systematic or deterministic effects from random or error effects [Zar, 1984]. Specifically, regression models assume that the probability distribution of a dependent variable y varies in a systematic fashion with an independent variable x. Typically, a continuous or discrete dependent variable is selected, and the value or occurrence probability of the dependent variable is calculated from a model that contains multiple independent variables. For example, in epidemiological medicine predictions of the lifetime prevalence of attempted suicide in the general community have been obtained using a weighted logistic regression model [Moscicki et al., 1988]. Specifically, such a model can contribute to an understanding of suicide and suicidal behavior in general populations. However, dynamical relationships, e.g., a difference equation, differential equation, or dynamical rule systems, are frequently not considered.

Recent developments in nonlinear prediction could provide techniques that may be able to predict complex phenomena in biological systems. For example, the prediction of occurrences of pathological changes in medicine and the folding of proteins is considered one of the foremost challenges. Currently, multiple research groups are working on the prediction of protein folding based on the primary sequence of proteins [Rost et al., 1993]. Moreover, predictive modeling of treatment responses to various medical conditions can be used to determine and improve efficacy of various treatment alternatives [Wiklund et al., 1992]. However, there appear to be conceptual gaps between the notion of predictability as it is used in nonlinear prediction and as it applies to biological systems. Thus, several aspects of predictability in biological systems will be reviewed briefly.

2 What to Predict

Predictions in biological systems that generate complex spatio-temporal evolutions can be divided into three major categories. First, the average value of a dependent variable within a predefined temporal interval is sought. For example, survival rates of pathological conditions such as cancer or cardiovascular disease are obtained via survival analysis for a fixed time period. Second, statistical models attempt to predict quantitative or qualitative measures of the dependent variable, such as sleep staging derived from EEG trajectories or tertiary structure of proteins derived from the amino acid sequence. Third, the occurrence probability of the dependent variable for a given temporal interval is predicted from the set of independent variables observed at an earlier time point. Fourth, the specific trajectory or sequence of a dependent variable is predicted. Thus, each of these types of prediction is motivated by different objectives and carries specific limitations. To understand these models, it is important to outline the purpose of prediction for biological systems.

3 Purpose of Prediction

The purpose of predictions in biological systems can be summarized briefly into five categories. First, predicting pathological states, e.g., seizures, cardiac arrythmias in medicine or defective protein folding in biology, may help to initiate interventions such as pharmacological manipulations or structural substitutions to prevent or minimize the occurrence probability of this state. In this case, useful predictive models should be based on meaningfully defined states that reflect the underlying biological mechanism. Second, predictions derived from an explicit mathematical model may provide a better understanding of the influence of the elements of the system on its spatio-temporal evolution. Specifically, dynamics of epidemiological systems have been said to depend crucially on the transmission rate, type, and mode as well as the overall occurrence rate within the population. Thus, a statistical mechanical model with quantitatively defined interactions between the individual elements will improve the understanding of how individual behavior relates to the overall epidemiological dynamics. Third, simulation of the biological system by the predictive model can generate experimentally testable hypotheses that relate distinct changes of independent variables to the outcome of the dependent variable. However, for these predictions to be experimentally meaningful, elements of the model have to correspond to experimentally testable variables. Fourth, the interactive adjustment of the model based on the discrepancy between predicted versus actual spatio-temporal evolution of the system can iteratively enhance the accuracy of the model. Finally, predicting trajectories of complex systems may help to suggest optimal adjustments of the environment to the system, e.g., epidemiological-economic planning in response to the dynamics of infectious diseases.

4 Prediction Models

Various predictive models have been developed with respect to the different types of predictive goals. One of the most frequently used models in medicine and epidemiology is derived from the logistic function. Here, the occurrence probability $P(y|x)$ of belonging to a category y, based on a set of independent variables, x, is given by

$$P(y|x) = \frac{e^{c+ax}}{1 + e^{c+ax}}. \tag{1}$$

The coefficients, c and a are estimated iteratively using the method of maximum likelihood [Dixon, 1989]. This model also allows for interaction between independent variables, i.e., cross-terms of the form $x_i x_j$. Another model that has been employed to predict time series data consists of auto-regressive moving average methods (ARMA). Briefly, an average of the dependent variable is obtained for a given time interval. In addition, the autocorrelation between different time points is obtained and the parameters for a linear combinations of these measures that minimize the least squares between actual and predicted values are estimated. The methods mentioned above represent examples of predetermined models for which parameters are fitted according to a particular metric. A different set of approaches are based on local rules of connectivity that evolve iteratively via statistical correction with respect to a fitting function. For example, different types of neural networks have been used for predictions for a variety of biological systems ranging from neuronal firing patterns [Amit and Treves, 1989] in the hippocampus to tertiary structure of proteins [Munson et al., 1994]. Similarly, genetic algorithms and classifier systems represent local rule-based systems that evolve to optimally match a particularly environment [Booker et al. 1989]. Lastly, a number of different geometric methods have been used to predict spatio-temporal evolution of experimental time series. These methods consist of dividing the phase space into subsets and using nearest neighbors of a data point to predict its successor [Sugihara and May, 1990].

5 Statistics of Prediction

An important aspect of statistical models as tools for prediction in biological systems is the question of whether the model is *generalizable*. Specifically, to what extent can a model that has been derived from a set of experimental data predict the spatio-temporal evolution of a different data set from the same biological system. Thus, it is important to quantify error rates of predictions when using the statistical models outlined above. However, there is no generally accepted method for obtaining prediction errors, since many error measures are defined with respect to statistical parameters that have to be estimated from the experimental data sets. For example, using a maximum likelihood estimation of a in a logistic regression model, the actual error rate can be defined via [Van-Houwelingen and Le-Cessie, 1990]

$$
\begin{aligned}
\mathrm{MML_{ACT}} = &- \frac{1}{n} \sum_i P(y|x_i) \log(P(y|x_i)_{\mathrm{estimated}}) \\
&+ (1 - P(y|x_i)) \log(1 - P(y|x_i)_{\mathrm{estimated}}).
\end{aligned}
\tag{2}
$$

Recently, data resampling methods such as cross-validation have been used to obtain estimates of $P(y|x_i)$. However, while cross-validation yields estimates of the probability that an observation belongs to a particular category, it does not indicate whether the logistic regression model selected to yield predictions about categorical membership is well-suited. In simpler categorical prediction models, odds ratios or chi-squared analyses have been used to determine the significance and appropriateness of the model. However, simulations of various statistical models that allow an explicit calculation of the actual error rates have shown that estimated error rates can be significantly different from actual error rates. Statistically, these results imply that, while a predictive model can be fitted to an experimental data set with high predictive power, the quantification of the extension of this model to other data sets may be difficult to obtain practically. In conclusion, further development of statistical methods will be required to determine whether a specific predictive model is generalizable to the spatio-temporal evolution produced by a specific biological system.

6 Limits of Prediction in Biological Systems

Limitations of prediction in biological systems are closely associated with the inherent characteristics of these systems. As mentioned above, there is ample evidence that many biological systems are neither completely deterministic nor stochastic, but that both necessity and chance crucially influence the spatio-temporal evolution. Apart from this theoretical consideration, there are many practical problems that occur when biological systems are observed experimentally. First, data sets in biology consist typically of small number of observations when compared to many physical experiments. For example, continuous time evolution systems such as EEG or ECG signals are recorded in batches of a few hundred to a few thousand data points. Moreover, observations of population characteristics, e.g. physiological or psychological parameters of a group of human subjects, comprise frequently only very few data points. The limited number of observations can be ascribed to several factors. Specifically, biological systems continuously adapt to their internal and external environment. However, these environments are changing on different time scales ranging from milliseconds to months. Simultaneously, the adaptation mechanisms available to biological systems also function on different time scales. For example, the NMDA (N-methyl-D-aspartate) class of glutamate receptor plays a critical role in a variety of forms of synaptic plasticity in the vertebrate central nervous system. One extensively studied example of plasticity is long-term potentiation (LTP), a remarkably long-lasting enhancement of synaptic efficiency induced in the hippocampus by brief, high-frequency stimulation of excitatory synapses. It has been proposed that LTP has two components acting on two different time scales. First, a short-term (millisecond), decremental component which can be mimicked by NMDA receptor activation, and a long-lasting (second to minute), non-decremental component which requires stimulation of presynaptic afferents [Kauer et al., 1988]. In humans the repeated use of amphetamine (AMPH) produces a hypersensitivity to the psychotogenic effects of AMPH that persists for months to years after the cessation of drug use. There is strong evidence that this behavioral sensitization is due to enhanced mesotelencephalic DA release, especially upon re-exposure to the drug [Robinson et al., 1988].

These examples emphasize that biological systems are highly non-equilibrium systems characterized by continuous changes on different spatial and temporal scales. Moreover, biologically relevant observations frequently extend over long periods of time relative to the time scales operating on the system at hand. For example, population characteristics or animal behavioral paradigms may yield relatively few observations collected during several months. Frequently, this limitation is compensated by observing many similar subjects and apply ensemble statistics to yield predictive models.

Second, biological systems consist of many elements that influence the spatio-temporal evolution. Typically, only few independent and dependent variables are considered for an experimental observation while the remaining variables are thought of as stochastic perturbation having minimal relevance for the experimental hypothesis at hand. While this approach successfully reduces the number of variables and renders a biological system experimentally treatable, the "stochastic" fluctuation of the observed variables are frequently found to be large relative to experiments with physical systems. Thus, the observed data sets are typically characterized by a low signal-to-noise ratio. While low signal-to-noise ratios permit simple statistical analyses that reveal the influence of various independent variables on the dependent measure, they severely limit the scope of predictive models.

To summarize, non-stationarity and low signal-to-noise ratio significantly curtail the temporal and spatial scaling range, respectively. The response to these limitations has been to limit the scope of prediction by categorizing continuous data sets or by predicting a state of the system characterized by a set of statistical measures. In addition, transformations of the original data set, including filtering, have been applied to increase the signal-to-noise ratio. Lastly, resampling methods such as "boot-strapping" or "jack-knifing" provide an alternative way to generate probability distributions for the observed data sets. The specific mathematical form of a model chosen to predict biological data remains arbitrary in general. With a few exceptions for which fundamental equations have been developed, e.g. the Hodgkin–Huxley equation for membrane potentials, most models are based on heuristic arguments. The a priori formulation of a specific model is often complicated even though the elements of the biological system under consideration are clearly discernible, e.g., specific neuronal systems in the central nervous system. The difficulties are due to the fact that the individual elements are highly interactive giving rise to a "latent variable" structure that transcends the physical entities of the system. Thus, it is not surprising that complex biological systems, e.g., coordinated hand movements, can be adequately described by a model with few parameters despite the fact that these systems comprise many individual elements. However, the lack of "first principles" and the organization of many highly interactive elements to form a globally characteristic spatio-temporal evolution, complicates the explanatory value of predictions generated by mathematical models with respect to the underlying biology.

Specifically, while simple models of complex biological systems can provide a general understanding of the different states of the system, they fall short of helping to relate the biologically relevant variables to the dynamics of the system's behavior.

Lastly, many biological systems exhibit sudden transitions from an experimentally well-defined state to a qualitatively different state. For example, the EEG can undergo dramatic changes from complex and possibly chaotic oscillations with different underlying major frequencies to the stereotypic oscillatory patterns defining different types of seizures [Gallez and Babloyantz, 1991]. Similarly, the ECG can change suddenly from its normal pattern to a highly irregular oscillation characterizing life-threatening arrythmias [Chialvo and Jalife, 1987]. In both cases, prediction models of these biological systems are challenged to undergo qualitative changes analogous to first-order phase transitions, in order to predict the onset of highly significant pathological events. Moreover, models of these systems should not only provide insights into the parameter constellations that promote these qualitative transitions but should also yield insights into the mechanisms of transition that enable preventative interventions.

7 Further Problems

In addition to factors that directly limit the prediction of biological systems, difficulties arise when multiple conceptual levels are used to describe the same system. For example, it has long been recognized that the overall "activation state" of the central nervous system can be influenced by a few neurotransmitter systems located in the mesencephalic portion of the brain [Robbins et al., 1989]. The different activation states have been labeled on an electrophysiological level according to the distinct EEG recordings associated with these states. Behaviorally, conceptual notions such as "attention" have been used to describe the ability to shift or maintain focus on a task at hand. Finally neurobiologically, differential activation of single norepinephrine-containing neurons in the locus ceruleus has been associated with sleep and wakefulness. Thus, an attempt to model and predict the overall "activation state" of the CNS will be confronted with concepts and mechanisms proposed on different levels of description. Moreover, while the connection between these levels has been difficult to establish experimentally, a theoretical connection may impose even greater challenges. Another problem that has been recognized concerns the fact that biological systems realize only a few states relative to the large number of states theoretically available to the system. For example, higher-order Markov analysis of nucleotide sequences of DNA shows highly repetitive sequences and small subsets of realized nucleotide sequences. However, the observed statistics of these occurrence probabilities are not only determined by the sparse realized set but also by the limited number of observations for any one DNA sequence.

8 Examples of Approaches to the Prediction of Biological Systems

8.1 Determining the Spatio-Temporal Scale for Optimal Prediction

Experimental time series in physics and biology result frequently from complex underlying systems that cannot be described easily by equations based on first principles. Nevertheless, consecutive experimental observations often exhibit hierarchical or sequential order properties that distinguish these observations from sequences of independent random variables. In particular, the correlation integral quantifies the average density of states, i.e., observations with similar spatio-temporal characteristics, in a reconstructed state space for various spatial resolutions. The density of these states reflects the spatio-temporal patterns obtained from the experimental observations. A uniform density of states is induced by an independent random process, i.e., the spatio-temporal patterns are solely due to the joint probability of different experimental observations. Consequently, a correlation integral deviating from a random process quantifies the association between experimental observations. Usually, the slope of the correlation integral with respect to spatial and temporal resolution is used to estimate the fractal dimension and a lower limit of the Kolmogorov entropy. However, the scaling range is frequently not sufficient for biological data to permit an appropriate fit. Instead, the correlation integral can be thought of as a local gauge function that quantifies the degree to which spatio-temporal structure changes the observed joint probabilities from the joint probabilities expected by chance. Specifically, the average density ρ, of two points within a distance r, given a total size of the reconstructed space of R in d-dimensional space, is given by

$$\rho \approx \left(\frac{r}{R}\right)^d . \tag{3}$$

By definition of an independent random process, the density of a sequence of two points $\rho_{1,2}$, to fall within a given radius r is given by the product of the densities for each point to fall within the radius r, i.e.,

$$\rho_{1,2} = \rho_1 \rho_2 . \tag{4}$$

Thus, given a sequence length or embedding l, the average density of points within a distance r is given by

$$\rho \approx \left(\frac{r}{R}\right)^l . \tag{5}$$

Taken together, the surface of the correlation integral for an independent random process is given by

$$\rho(l, d) \approx \rho_0^{l \times d} . \tag{6}$$

Consequently, the decay of the density is linear in d and l along the d and l axis, respectively, and geometric in between. Thus, by comparing the correlation integral surface (CIS) of an experimentally observed data set with the same data set of randomized order, information is obtained of the structure contained in the data set. Moreover, the values of the correlation integral difference surface (CIDS) for different values of r and l indicate the spatial and temporal scale that most distinctly separates the observed system from the randomized system. Moreover, both CIS and CIDS can be used for statistical comparisons of experimental manipulations [Paulus et al., 1993]. These comparisons involve either the generation of surrogate data based on some randomization method [Knuth, 1968] or resampling and boot-strapping methods [Efron and Gong, 1983] to obtain estimates of errors for these measures. Lastly, the interpretation of the generalized correlation integral as measuring the average density of states leads to the embedding of this tool into the realm of statistical mechanics. Specifically, the interpretation of the generalized correlation integral as a canonical partition function as proposed by [Badii, 1989] enables us to interpret qualitative changes of this function as due to phase-transition-like behavior in the experimental data. This notion appears to be of particular importance for some biological systems that exhibit switching between qualitatively different states.

As an example, the CIS and CIDS is shown for a rat self-administering cocaine via an intravenous catheter. Briefly, albino Wistar rats (400-450 g) were prepared with intravenous jugular catheters and trained to self-administer cocaine hydrochloride (0.25 mg/injection) on a fixed-ratio 5 reinforcement schedule as described by [Markou, 1991]. After a training phase of daily 3 h SA sessions in which the behavior stabilized, animals were allowed to self-administer cocaine for 48 h continuously. Both the number of self-administered cocaine injections and the inter-reinforcement intervals (IRIs) were collected. The Fig. 1 shows a three-dimensional graph of CIS and CIDS for an exemplary animal. The x-axis corresponds to the sequence length, the y-axis indicates the logarithm of the distance and thus signifies the spatial resolution, and the z-axis displays the logarithm of the probability of similar IRI sequences (CIS) or the logarithmic likelihood ratio between original and randomized data (CIDS). The CIS shows the expected decay of the probability of similar sequences with increasing sequence length as well as with increasing resolution. The log-likelihood ratio given by the CIDS indicates that for a low level of spatial resolution (between 2^{-2} and 2^{-4} of the standard deviation of the data) and sequence length as long as 10, significant differences exist between the observed probabilities and the expected chance probabilities. Thus, sequences of self-administration inter-reinforcement intervals exhibit significant nonlinear structure on low levels of resolution for extended sequence lengths. These results have important implications for predictive models. It appears that, instead of trying to predict the precise inter-reinforcement interval, models should be developed that predict sequences of "short" versus

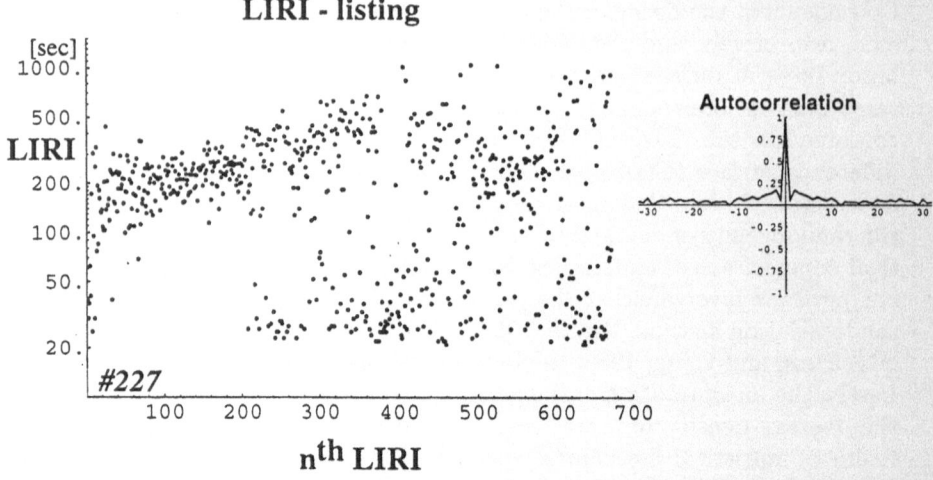

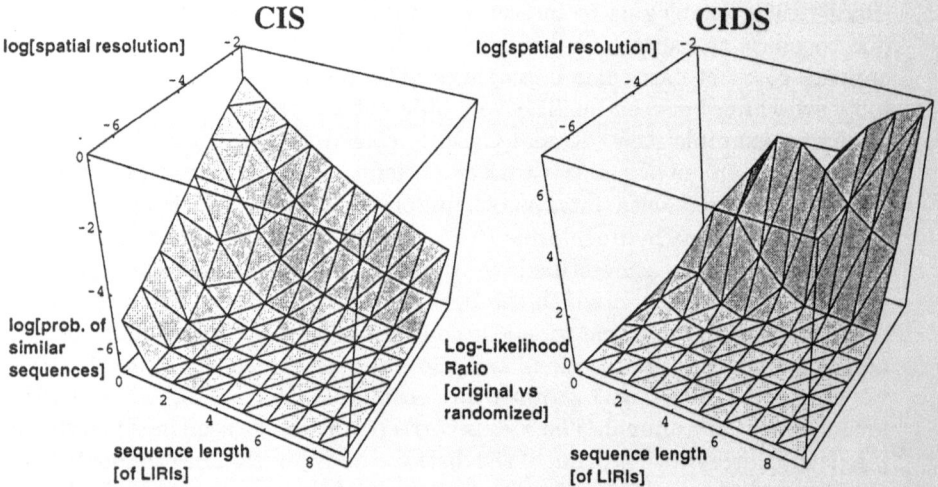

Fig. 1 A three-dimensional graph of CIS and CIDS for an exemplary animal (see explanations in the text).

"long" inter-reinforcement intervals. Moreover, the distant "past", i.e., the last 10 inter-reinforcement intervals, significantly influences the duration of the next interval despite the fact that the autocorrelation function falls off rapidly (Fig. 1).

8.2 An Order Parameter Model for Behavioral Organization

Behavioral organization can be defined as the selection, ordering, and sequencing of behavioral elements in response to external or internal stimuli to form flexible yet stable macroscopic patterns of behavior. The emerging behavioral patterns result from the interaction of actions, plans, and observations. For example, a person waiting at the side of a road (action) to cross the street (plan) needs to monitor the traffic and watch the traffic light (observation). Subsequently, the person will wait or initiate movements, monitor the traffic while crossing the street, and compare the current actions with the underlying plan (e.g., to reach a certain location). These actions require a sophisticated organization of the available responses and concentration on selective sensory and cognitive elements. Organizational principles can be defined as the rules that determine the arrangement of behavioral elements into interdependent parts to form behavioral patterns. It is assumed that these rules can be explicated as possibly nonlinear functions between behavioral sequence characteristics and measures of the interdependence of behavioral elements. Specifically, it is hypothesized that a set of order parameters (macroscopic variables that determine behavioral patterns) can be extracted from experimental data sets of human or animal response sequences. A key to the quantitative analysis of behavioral organization is the assessment of the interdependence between behavioral elements or the measurement of the degree of association between consecutive behavioral elements. The dynamical entropy measure [Bohr and Rand, 1987] has been developed based on the ergodic theory of dynamical systems [Ruelle, 1978] to quantify the change in information about a dynamical system by observation of its time series (trajectory). A technique analogous to nearest neighbor methods has been suggested to calculate efficiently these local dynamical entropies [Grassberger, 1989]. This technique is based on the idea that the uniqueness of a sub-sequence within a data sequence is related to the degree of association between its elements. A sub-sequence with a low degree of association between its elements can be identified by a short unique sub-sequence length. In contrast, highly interdependent elements yield a subsequence that is identified by a long unique subsequence length. Thus, the local dynamical entropy is large for the former and small for the latter. Additional sequence characteristics can be quantified by order parameters in analogy to statistical mechanics [Chandler, 1987]. In physical systems, a locally defined quantity such as the magnetic moment is used to obtain a control parameter-dependent function of the overall magnetization of the system. In this series of experiments, subjects were asked to choose between two response alternatives (pushing the left button = "0", pushing the right button = "1"). For each subject, a sequence of 500 consecutive choices was recorded on a computer. For this binary response sequence, a "switch response", i.e., "01" or "10", is denoted by a 1, whereas a "stay response", i.e., "00" or "11", is assigned to 0. In addition, the following order parameters can be obtained: response balance (difference between left and right choices); inter-response interval; and response location. The order parameter approach was developed to extract the macroscopic properties of a complex system by inferring the microscopic structure of the underlying process [Haken, 1989]. This approach is based on the maximum entropy principle and can be applied to both equilibrium and non-equilibrium systems by choosing adequate constraints. These constraints are selected from functions of the macroscopic variables of the experimental system. Since the macroscopic behavior of physical systems is determined crucially by the potential function describing different energy states as a function of physical parameters, the main objective of this approach is to obtain a functional description of the potential of a variable or a set of variables. The link between model and experiment is the fluctuation spectrum of local dynamical entropies, $S(h)$ [Bohr and Rand, 1987]. The order parameter model provides a procedure to derive an explicit partition function, which is used to calculate $S(h)$. The following order parameters were used for the initial model development: *response switching* (s), *response balance* (b), and *response duration* (d). Central to the functional development is the computation of local dynamical entropies via $h_i = \log_2 N/l_i$. Thus, the sequence length, l_i which uniquely defines the subsequence starting at the ith position is the measurable variable that determines the degree of association between behavioral elements. The model development attempts to identify a function that defines l_i in terms of response switching, s, response balance, b, and response duration, d, by a set of order parameters, op, i.e., $l = f(s, b, d, op)$. For the development of a model describing the fluctuation spectrum of local dynamical entropies, the data of 15 control subjects were pooled. Each data set consists of 500 unique subsequences characterized by a specific sequence length, l, response switches, balance, and average response duration. Consequently, 7500 data points were used to estimate the order parameters for a second-order polynomial. Specifically, the following coefficients were fitted using a stepwise regression procedure [Dixon, 1989]: the switching, sw, balance, ba, duration, du, squared switching, sqs, squared balance, sqb, squared duration, sqd, switching-balance, $swba$, switching-duration, $swdu$, and balance-duration, $badu$.

Table 1.

VAR.	COEFF.	SE(COEFF)	TOL.	F
CONST.	21.4594			
sw	−35.9306	0.6663	0.0583	2907.56
ba	1.5245	0.1746	0.2773	76.24
du	$-0.534E - 02$	$0.23E - 03$	0.1293	535.75
sqs	26.8002	0.703	0.0612	1451.01
sqb	−0.9124	0.2136	0.7559	18.24
$swba$	−2.8611	0.4990	0.2823	32.87
$swdu$	$0.976E - 02$	$0.47E - 03$	0.1002	417.45

The fitted model consists of 7 parameters and explains 43% of the variance (Table 1). The switching, squared switching, and switch-duration coefficients contribute most significantly to the fitted model. The tolerance values

(TOL) for most coefficients indicate that some of the independent variables are collinear. Hence, the model parameters could be extracted from the principal components of the independent variables, an approach that has also been suggested by [Haken, 1989]. However, it may be difficult to interpret these principal components experimentally. A preliminary analysis using a stepwise regression based on two principal components yielded a model of the l function that explained only 32% of the variance.

The graph of the fitted function for various combinations of the order parameters provides important insights into the principles governing behavioral organization in this paradigm (Fig. 2). Specifically, the function of the unique subsequence length, l, is shown for the combinations of switching and balance, switching and duration, as well as balance and duration, respectively. These graphs indicate that the unique sequence length is a strongly nonlinear function of response switching and response balance, as well as response switching and response duration. In particular, the unique sequence length, l, determining the degree of association between consecutive response elements, is lowest for a probability of 0.5 for response switching and increases for both low and high probabilities of response switching (left). Moreover, the unique subsequence length decreases with low probability of switching for longer inter-response intervals but increases for a high probability of switching and longer inter-response intervals (middle). Lastly, the association between response elements decreases monotonically with increasing inter-response interval regardless of the response balance (right).

9 Challenges for the Development of Prediction Models in Biology

From the notions and problems outlined above and the two approaches briefly described by the examples, several challenges emerge for models of prediction of biological systems. First, the limited number of data points requires statistical and analytical methods that extract the maximum information from the data set. Currently, this could strictly be defined as the difference between an assumed underlying random process and the amount of structure reflected in the observed joint probabilities of sequences of observation on a parameterized spatial scale. Moreover, statistically, resampling methods could be used to estimate statistical distributions underlying the observed data points. Second, further developments are required of methods that determine the optimal spatio-temporal scale for a model. Currently, models of physical systems are frequently obtained on the highest spatial resolution possible. However, biological systems may not provide a sufficient signal-to-noise ratio to generate a high-resolution model. Moreover, biologically "relevant" information may not be contained on high-resolution scales. Third, statistical methods need to be developed to quantify the validity of the model and the extent to

Choice Task: 15 Control Subjects, Step-wise Regression Fit

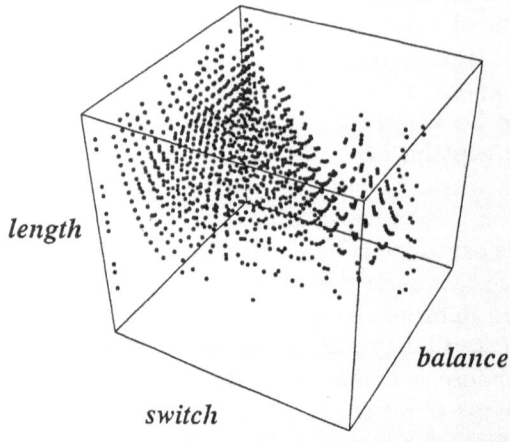

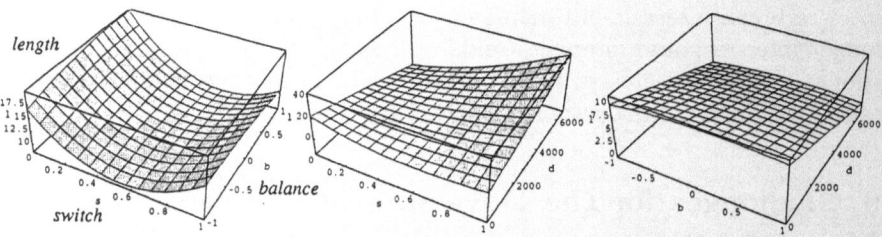

Fig. 2 The graph of the fitted function for various combinations of the order parameters.

which it can be generalized. Split sample approaches, cross-validation, and other re-sampling methods provide procedures to test the parameters of a predictive model. Fourth, predictive models have to be incorporated into the experimental and a theoretical framework. Specifically, the parameters of a model have to be experimentally modifiable to help provide insights into the underlying biological mechanisms. In addition, to understand the system's behavior theoretically, concepts that have been used to describe complex physical systems may lead to useful descriptors of the overall state of a biological system. Thus, embedding of predictive models into a synergetic or statistical mechanics framework may yield a better understanding of global changes and qualitative transitions in biological systems. In conclusion, predictability of biological systems is challenged by difficult problems inherent in the nature of these systems; however, development of methods that can extract predictive models will significantly improve our understanding of these systems.

References

Amit, D.J., Treves, A. (1989): Associative memory neural network with low temporal spiking rates. Proceedings of the National Academy of Sciences of the United States of America **86**(20), 7871–7875.

Badii, R. (1989): Conservation laws and thermodynamic formalism for dissipative dynamical systems. Rivista Del Nuova Cimento **12**, 1–72.

Bohr, T., Rand, D. (1987): The entropy function for characteristic exponents. Physica **25D**, 387–398.

Booker, L.B., Goldberg, D.E., Holland, J.H. (1989): Classifier systems and genetic algorithms. Artificial Intelligence. **40**(1-3), 235–282.

Chandler, D. (1987): Introduction to Modern Statistical Mechanics. Oxford University Press.

Chialvo, D.R., Jalife, J. (1987): Non-linear dynamics of cardiac excitation and impulse propagation. Nature **330**(6150), 749–752.

Dixon, W.J. (1989): BMDP Biomedical Computer Programs. Los Angeles: University of California Press.

Doyon, B. (1992): On the existence and the role of chaotic processes in the nervous system. Acta Biotheoretica **40**(2-3), 113–119.

Efron, B., Gong, G. (1983): A leisurely look at the bootstrap, the jackknife, and cross-validation. The Amer. Stat. **37**, 36–48.

Freeman, W.J., Skarda, CA. (1985): Spatial EEG patterns, non-linear dynamics and perception: the neo-Sherringtonian view. Brain Research **357**(3), 147–75.

Gallez, D., Babloyantz, A. (1991): Predictability of human EEG: a dynamical approach. Biological Cybernetics **64**(5), 381–391.

Grassberger, P. (1989): Estimating the information content of symbol sequences and efficient codes. IEEE Trans Inform. Theory **35**, 669–675.

Haken, H. (1983): Synergetics: An Introduction. Springer, Berlin.

Haken, H. (1988): Information and self-organization : A macroscopic approach to complex systems. Springer, Berlin New York.

Kauer, J.A., Malenka, R.C., Nicoll, R.A. (1988): NMDA application potentiates synaptic transmission in the hippocampus. Nature **334**(6179), 250–252.

Knuth, D.E. (1968): The art of computer programming. Addison-Wesley, Reading, Mass.

Markou, A., Koob, G.F. (1991): Postcocaine anhedonia: An animal model of cocaine withdrawal. Neuropsychopharmacology **4**, 17–26.

Moscicki, E.K., O'Carroll, P., Rae, D.S., Locke, B.Z., Roy, A., Regier, D.A., (1988): Suicide attempts in the Epidemiologic Catchment Area Study. Yale Journal of Biology and Medicine **61**(3), 259–268.

Munson, P.J., Di Francesco, V., Porrelli, R. (1994): Protein secondary structure prediction using periodic-quadratic-logistic models: statistical and theoretical issues. in: Proceedings of the Twenty-Seventh Hawaii International Conference on System Sciences. Vol.V: Biotechnology Computing (Cat. No.94TH0607-2). Edited by: Hunter, L. Los Alamitos, CA, USA: IEEE Comput. Soc. Press. 375–384.

Paulus, M.P., Kadtke, J.B., and Menkello, F. (1993): Statistical mechanics of biological time series: assessing geometric and dynamical properties. International Journal of Bifurcation and Chaos **3**, 712.

Robbins, T.W., Cador, M., Taylor, J.R., Everitt, B.J. (1989): Limbic-striatal interactions in reward-related processes. Neuroscience and Biobehavioral Reviews 13(2-3), 155–162.

Robinson, T.E., Jurson, P.A., Bennett, J.A., Bentgen, K.M. (1988): Persistent sensitization of dopamine neurotransmission in ventral striatum (nucleus accumbens) produced by prior experience with (+)-amphetamine: a microdialysis study in freely moving rats. Brain Research 462(2), 211–222.

Rost, B., Schneider, R., Sander, C. (1993): Progress in protein structure prediction? Trends-Biochem-Sci. 18(4), 120–123.

Ruelle, D. (1978): Statistical Mechanics, Thermodynamic Formalism. Addison-Wesley, Reading Mass.

Schaffer, W.M. (1985): Can nonlinear dynamics elucidate mechanisms in ecology and epidemiology? Ima. J. Math. Appl. Med. Biol. 2(4), 221–252.

Sugihara, G., May, R.M. (1990): Nonlinear forecasting as a way of distinguishing chaos from measurement error in time series. Nature 344(6268), 734–741.

Van-Houwelingen, J.C., Le-Cessie, S. (1990): Predictive value of statistical models. Stat-Med. 9(11), 1303–1325.

Wiklund, K., Hakulinen, T., Sparen, P. (1992): Prediction of cancer mortality in the Nordic countries in 2005: effects of various interventions. Eur. J. Cancer. Prev. 1(3), 247–258.

Zar, J.H. (1984): Biostatistical Analysis. Prentice-Hall, Englewood Cliffs.

Limits of Predictability for Biospheric Processes

Nikita N. Moiseev

Abstract

Understanding the possible course of physical processes, and the repercussions of man's actions, has certain natural limits. Knowing these is no less important than being able to foresee the future. The present work deals with these issues within the context of the problem of the biosphere stability.

1 Methodological Premises

In this work, I shall adhere to one of the basic principles of modern rationalism stated by Niels Bohr. According to Bohr: "Only those things exist that can be observed". This principle allows us to visualize the surrounding world, the Universe, as a single (whole) system. Therefore, all the processes occurring within it should be looked upon as sequences of functioning of some self-organizing systems. This carries over to rational activity, which we shall take as a natural phenomenon, that is, belonging to the Universe.

Conceptions of the Universe's nature other than those fitting with Bohr's principle make no sense. Indeed, if the Universe were not a system, how could we have discovered this? From such a standpoint, the assumption of the Universe as a system is quite trivial. However, what follows from it is directly related to our subject of inquiry. First and foremost, the observer (researcher) is, by our assumption, an element of a system, as is the subject of observation. Moreover, they are connected with one another and with the rest of the system by bonds that can in principle be traced down (say, gravitational forces). Thus, to discriminate a subject of observation (including processes evolving with time) appears to be a non-trivial operation. Locating it is only possible if there is some finite time interval during which the effect of the system (and hence, of the observer) can be regarded as external forcing. Which means that from the point of view of the observer, changes in the system's properties (and thus, in the effect it produces upon the observer) can be neglected by virtue of the changing condition of the object and the observer's actions.

Such a premise of rationalism, an empirical "generalization", as Vernadsky would say, puts obvious limits not only predictability, but also the conception of Truth.

The evolution of the Universe itself, along with the processes occurring within it, should be understood as self-organization (or universal evolutionism), since in the context of Bohr's principle there are no external forces, or causes, governing this evolution. The element of particular value for the Universe's functioning is Reason. According to a rationalism paradigm, Reason, along with the living matter, are natural phenomena that came into being at a certain stage of the Universe's evolution. In other words, a time come when the Universe, at least part of it represented by *homo sapiens*, had acquired the tools for self-knowing – the intellect of individuals and the collective intellect of mankind. Those possessing these tools have certain cognitive advantages enabling them to look into the future. The capabilities of Reason as a natural phenomenon and part of the Universe system are limited and determined by the level of evolution. Therefore, all processes involving human beings are directly concerned with self-organization, and their future depends not only upon the evolutionary history and random perturbations that are inevitable, but also on the predictability horizon and other properties of Reason as an element of the system. Naturally, the intellect develops continuously. Whereas morphologically we, the moderns, are practically undistinguishable from our Cro-Magnon ancestors, and whereas the human brain, and thus, intellectual potentialities have virtually not changed since the times of mammoth hunters, the collective intellect has been increasingly developing during the ages that elapsed. As a result, our ability to cognize the world around us has dramatically grown, particularly during the last century.

How far is this going to proceed? The process may be restricted both physiologically and by the social organizational structure. Furthermore, the evolution of collective intellect spurs the development of both science and technology and productive forces. This in turn speeds up evolutionary changes in Nature and society. The paradoxal result is a reduction in of the extension of the predictability horizon.

Let me once again cite Niels Bohr, namely, his widely-known statement that no complex phenomenon can be described using one language. Classical rationalists of the 18th and 19th centuries contented themselves with the concepts of absolute determinism to describe the laws of nature. This, however, proved insufficient: the description of the laws of physics and other self-organization phenomena requires the language of stochastics. Besides, the role of uncertainty must be taken into account. As quantum physics has shown, these requirements come from the very nature of the Universe. At the macro-world level, we can only manipulate with probabilistic distribution.

Stochasticity and uncertainty penetrate all of the Universe's structure. Mutagenesis, different social phenomena and virtually all biospheric processes are only some of the possible examples. The American meteorologist Lorenz has demonstrated that the atmosphere as a system has a limited "memory" of the order of 2 – 3 weeks. This means that the predictive power of the atmospheric circulation forecasts based on equations of the atmosphere dynamics is no more than $2 - 3$ weeks.

It is the social life of mankind where uncertainty and stochasticity exhibit themselves in a particularly pronounced manner. The reason for this, among other things, is a striking diversity of human individualities. Any human being is a personality that perceives and assesses events, makes decisions and acts in his (her) own way.

As a consequence, it is impossible to predict details of practically any natural process, especially one involving man. Thus, it becomes a matter of special interest to study adiabatic invariants, that is, the integral characteristics of the process under review whose rates of change are smaller than typical rates of other characteristics of a process and whose calculation involves averaging in one way or another over the probabilistic component. The latter point is equally true for both natural processes and those of social origin.

An important role in the evolution process is played by the mechanisms of bifurcation (polyfurcation, or catastrophes, as Rene Thom would say). The first study of the bifurcation phenomenon is due to Leonhard Euler, who treated the buckling of a column subjected to axial compression. He discovered that once a certain critical load value is reached, the column loses its vertical stability and under the action of, say, an occasional gust of wind it starts buckling about a new equilibrium position. What is more, there appears a continuum of new equilibria, comprising a surface formed by a rotating sinusoidal half-wave. Thus, it is impossible in principle to predict about which new equilibria the column will buckle, since the process is initially dependent upon the particular gust of wind that had occurred at the moment when the column had ceased being straight in equilibrium. At that moment, a failure had occurred in the system's memory.

The situation investigated by Euler as early as the 18th century is just a refined example of how the character of a dynamical process can be shifted. In most cases, however, we deal with bifurcations that are not momentary events, and the memory of previous conditions does not vanish entirely. Once the condition of neutral equilibrium is reached, a system starts restructuring itself, in a manner quite unpredictable, since random factors begin to assume a significance, and a variety of "channels of evolution" becomes possible.

The examples of bifurcations in social life are revolutions and rapid restructurings. I can cite no historical example of a revolutionary restructuring in which initial goals and objectives have been implemented.

Thus, a bifurcation is a condition setting natural limits to predictability. However, if a system's parameters change quickly enough, the impact of noise factors decreases, which opens up new possibilities for predictions within systems undergoing bifurcations. This conforms to the principle: The higher the rate of change, the easier it is to predict [see the contribution by Butkovskii, Kravtsov and Brush in this book].

Whereas bifurcation points are among the causes of uncertainty setting limits to predictability, attractors may allow one to determine whether a system will actually behave that way.

The remarkable property of dynamical systems is that in certain regions of phase space their trajectories (conditions, or solutions) may attract (or repel, if the attractor is unstable) other trajectories of a given system. Hence, once we know about the existence of an attractor in a process under study, we can state with assurance that this process will develop in such and such way.

In studying the processes that take place in both inert and living matter and in society, we run into quite a variety of attractors. Of special significance among them are the so-called strange attractors. A striking thing about them is that when emerging along a dynamical process trajectory they may show up not as specific conditions (points in the phase space) or trajectories, but as phase space regions filled with the process trajectories. The trajectories inside such a region have the property of local instability making a perturbation grow exponentially. As a result, predictions about the trajectories inside a strange attractor are only possible for limited lengths of time. In such instances, one says that a system has a narrow (or short) predictability horizon [Kravtsov, 1989, 1993].

Among the problems concerned with determination of future behavior and its susceptibility to understanding, are numerous problems of how to define and estimate stability of a dynamical system, and how to retain stability of particular properties over a certain span of time. The question of defining stability is of central importance to the analysis of prediction problems. The investigator may be concerned with quite a variety of characteristics. Therefore, there is still no clear and all-embracing definition of stability, the notion nowadays is increasingly replaced by a no less indeterminate term "sustainability". In each individual case it can bear a specific meaning, from classical stability in the sense of Lyapunov to exotic negative feedbacks maintaining the efficiency of certain properties of a system.

In analyzing these particularities, we sometimes come up against practically unsolvable problems. Let us see how this happens.

2 The Problem of Biospheric Stability

Preservation of the biospheric stability under the growing anthropogenic pressure is of fundamental importance to mankind. Even small changes of the biospheric characteristics may involve dramatic repercussions. Yet, some of these characteristics have already started to change due to man's activities.

Until recently, the role of a regulator controlling the state of the biosphere within certain limits has been played by the biota. Regulatory mechanisms of living matter are quite effective. They are capable of realization of various negative feedbacks, and over billions of years the living matter (mainly, plant life) has managed to compensate for externally induced perturbations of volcanic, cosmic or other origin. However, there is good reason to believe that the biospheric compensatory abilities may prove insufficient to resist the increasing anthropogenic pressure. As a result, the biospheric properties may get out of those stringent limits, within which a civilization can evolve.

That is why the ability to predict (estimate) changes in biospheric parameters under the increasing anthropogenic pressure is a relevant problem of modern science – and one of the most complicated, too. The possibilities to resolve it with reasonable accuracy are limited. As will be shown below, the very statement of the problem of stability is intricate. Besides, it is not easy at all to explain what "stability" means in this context. With "sustainability", the term increasingly applied to natural and social events, it is necessary to specify the meaning in each individual case. Now, some problems of a fundamental nature that we face in tackling the stability problem will be discussed. We can demonstrate them with the example of assessing the impact of growing carbon dioxide concentration in the atmosphere.

Let us consider a simple model of the interplay between the body of phytomass G and carbon dioxide concentration p. Denote the flow of carbon dioxide, i.e., the amount entering the atmosphere in unit time, by P. Then, the following balance equation can be writen:

$$dp/dt = cP - b_p p - \phi(G). \tag{1}$$

Here c is some constant, the term $b_p p$ gives the up-take of carbon dioxide by the ocean, the coefficient b_p is positive at temperatures below some critical value T_1. The term $\phi(G)$ describes the up-take of carbon dioxide by the biota (primarily, by plants). The function $\phi(G)$ is a concave one and can be well approximated by a power dependence G^s, where $s < 1$.

The change in phytomass G is given by a simple demographic equation

$$dG/dt = F(p)G - fG. \tag{2}$$

The "birth rate" $F(p)$ is a monotonically increasing function of concentration p. The "death rate" f is assumed to be constant. Under the above assumptions, the set of equations (1), (2) has a single stationary solution (p, G), defined by the equations:

$$cP = b_p \bar{p} + \phi(\bar{G}), \qquad F(\bar{p}) = f. \tag{3}$$

They make possible the estimation of variations in the carbon dioxide concentration and in the phytomass body according to the carbon dioxide flow entering the atmosphere, given that the flow rate is constant. A simple question now comes up: what will happen if the flow P changes significantly? Will p and G assume new stationary values, determined by (3)? Clearly, we can arrive at an answer after studying the stability of stationary solutions to the set of equations (1), (2). To this end, let us assume that

$$p = \bar{p} + y , \qquad G = \bar{G} + z .$$ (4)

By substituting (4) into (1) and (2), using (3) and saving only left-hand-side terms we get:

$$dy/dt = -b_p y - b_g z , \qquad dz/dt = ey ,$$ (5)

where

$$b_g = \left(\frac{d\phi}{dG} \right)_{G = \tilde{G}} , \quad e = \left(\frac{d\phi}{dp} \right)_{p = \tilde{p}} ,$$

or

$$\frac{d^2 z}{dt^2} + b_p \frac{dz}{dt} + eb_g z = 0 .$$ (6)

Equation (6) describes damped oscillations of a simple pendulum, and it follows from it that the stationary solution

$$\bar{G} = \bar{G}(P) , \qquad \bar{P} = \bar{G}(P)$$ (7)

is asymptotically stable.

Therefore, in the event of sudden ejection of carbon dioxide causing no appreciable climatic changes, the biota will be able to compensate for this effect by aggregate increase of the biomass, so that the "biota + atmosphere + carbon dioxide" system will remain in equilibrium.

This example has, perhaps, a more general meaning, since it demonstrates that for those models where stationary solutions are possible, sustainability can be treated classically as stability. Now, if the new value of carbon dioxide flow density does not differ much from the old one, the formulas (7) enable information of practical significance be obtained, and can be used to predict future behavior. However, the prediction made using these formulas has sense only as long as changes in carbon dioxide concentration do not entail temperature variations, that is, provided the coefficient b_p can be considered constant.

Note, that Le Chatelier's principle, in its classical sense, is realized in the "phytomass + carbon dioxide" system. In this instance, it arises from the laws of conservation written in the form of equations (1), (2).

The example considered in the previous section simulates an abrupt change in the carbon dioxide flow. Of even more interest is the situation where the value P grows with time as a result of anthropogenic emissions:

$$P = P(\varepsilon t) ,$$

where ε is some small parameter representing a slow process.

It is no easy matter to apply the concept of stability. Moreover, it is difficult to choose an "unperturbed solution", to say nothing of yet undeveloped methods for investigating stability of non-stationary motions. Fortunately, to this case one can apply the technique for analyzing solutions to problems of singular perturbation, based on Tikhonov's theorem [Tikhonov, 1952].

Let us substitute the independent variable $t = \tau/\varepsilon$ in the system of equations (1), (2). Then, (1) and (2) can be rewritten in the following way:

$$\varepsilon\frac{dp}{d\tau} = P(\tau) - b_p p - \phi(G) , \quad \varepsilon\frac{dG}{d\tau} = F(p)G - fG . \tag{8}$$

Tikhonov's theorem supposes that for Eq. (8) the solution to the Cauchy problem with the initial condition

$$p(0) = p_0, \qquad G(0) = G_0$$

for the time interval $t^* = O(1/\varepsilon)$ can be written in the form of the sum:

$$p = p^* + y + O(\varepsilon) , \qquad G = G^* + z + O(\varepsilon) , \tag{9}$$

where y and z are asymptotically stable solutions to the system of equations (5), while p^* and G^* are a separate solution to the system of equations:

$$P(\tau) - b_p p - \phi(G) = 0 , \qquad F(p) - fG = 0 .$$

It is suggested that values of $y(0)$ and $z(0)$ differ little from those of $p(0) - p^*(0)$ and $G(0) - G^*(0)$.

In our case, these conditions seem to be met. The p^* and G^* are functions of time growing monotonically, given the "external forcing" features the same quality. A qualitative behavior pattern of the initial system solutions is shown in Fig. 1. Here y and z are called "boundary layer" functions.

In the situation under review, one can forecast the character of carbon dioxide concentration changes only for times shorter than $t^* = O(1/\varepsilon)$ (see Fig. 1), and only under the condition that temperature changes related to the greenhouse effect are ignored. If these conditions are met, we can state that the biota retains its compensatory properties over this time interval and that the "carbon dioxide + biota" system is stable.

All the above considerations started from the assumption that the planet's mean temperature was stable. In actual fact, it varies, due both to the greenhouse effect caused by increased concentrations of carbon dioxide (and some other gases) and to the mean albedo varying according to vegetative and glacial cover. Hence, for a more detailed analysis one has to augment the model to include the climatic aspect. Quite suitable for this purpose and simple enough is Budyko's climatic model [Budyko, 1980] represented by the equation

$$dT/dt = \frac{S}{4}(1 - \alpha) - A - BT , \tag{10}$$

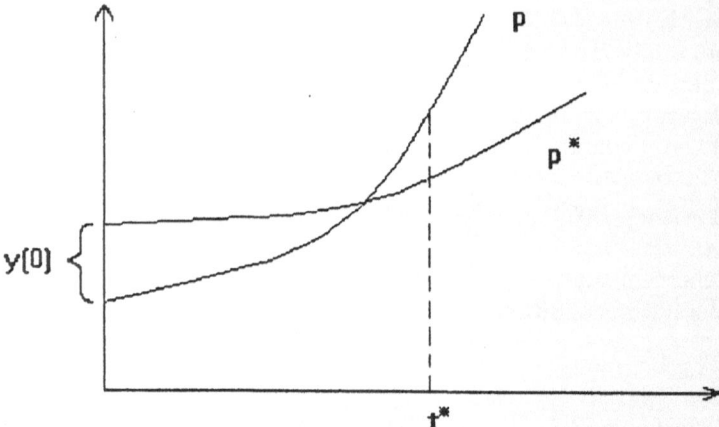

Fig. 1 The quasi-equilibrium solution p^* and the complete solution p as qualitative functions of time. Beyond the time interval $t^* = O(1/\varepsilon)$, the approximation (9) becomes invalid.

where T is the mean atmospheric temperature, S is the average flow of solar energy, α the mean albedo and A and B are some empirical coefficients originally defined by Budyko in 1962. The quantity $A + BT$ denotes heat radiation of the planet, in circumstances where the deviation from the equilibrium state is fairly small. Coefficients A and B are functions of carbon dioxide and other greenhouse gas concentrations and of cloud cover. The albedo α depends on type of vegetation and, generally, is a slowly increasing function of the biomass body G.

To account for the influence of varying temperature, it is necessary to consider an extended system "carbon dioxide + phytomass + atmosphere mean temperature". In the simplest case, it is described by equations (1), (2) and (10). These coefficient b_p denoting carbon dioxide up-take by the ocean is a decreasing function, and at a certain temperature $T = T_1$ it reverses sign.

Taking $P(\tau)$ as a slowly increasing function of time (since "anthropogenic" carbon dioxide enters the atmosphere in increasing quantities) and using the above formalization, we get

$$T = T^*(\epsilon t) + x + O(\epsilon) \,,$$

$$p = p^*(\epsilon t) + y + O(\epsilon) \,, \tag{11}$$

$$G = G^*(\epsilon t) + z + O(\epsilon) \,,$$

where T^*, p^* and G^* represent quasi-equilibrium behavior, while x, y and z are boundary layer functions. Notice that on the strength of (10), T^* is a slowly and monotonically increasing function of time, since p^* features the same property:

$$p^* = \frac{1}{b_p} [P(\varepsilon t) - \phi(G)] \,. \tag{12}$$

Similarly, A and B are monotonically decreasing functions of p^*, and hence, of time.

As the mean temperature increases, the coefficient of carbon dioxide take-up by the ocean goes down. At a certain value $T = T_1$ it damps down to zero, then reverses sign. As a consequence, the "carbon dioxide + biota" system, in conformity with (6), starts behaving as a negatively damped pendulum, and the set (11) loses its significance. Correspondingly, the "atmosphere + ocean + biota" system ceases to be sustainable, and its behavior changes qualitatively.

Complication of the models of biospheric processes due to inclusion of new factors affects our estimations, their reliability, and thus, the limits of predictability. Depending on the level of formalization, that is, on how detailed a description is, all characteristics of the evolution trajectory, including stability (sustainability) may significantly change, because a system's description becomes more complicated with augmentation of a model.

Under these circumstances, representation of a process as a singularly perturbed one eases the use of the term "stability" in its traditional sense and fairly accurately reflects a system's properties. Indeed, it makes possible estimation of the time interval over which a non-stationary quasi-equilibrium behavior becomes stable.

But the extended representation of stability (as sustainability) lacks generality and is far from always applicable to predicting process development, even when well-formalized. A case in point is the investigation of the interplay between the biota and climatic factors, when the employment of primitive, so-called zero-dimensional models failed to provide necessary information, and because of this general equations of hydrodynamics of the atmosphere and the ocean were called for. In order to construct a computer model on the basis of these equations we need their finite-dimensional analogs. To this end, one can use either Fourier–Galerkin representations (a spectral model) or finite-difference approximations to spatial variables. Either way the problem is reduced to that of a multi-dimensional model of the form

$$d\boldsymbol{x}/dt = X[\boldsymbol{x},\ P(\varepsilon t)] \,, \tag{13}$$

where $\boldsymbol{x}$ is a vector representing the system's condition. It can be either a coefficient of Fourier expansion within a spectral system or the value of a hydrodynamical quantity found at nodal points of a grid. P in (13) gives the vector of anthropogenic pressure. To apply the theory and method of singular perturbations, we need to know stationary conditions that are solutions to the following system:

$$X(\boldsymbol{x}, P) = 0 \,. \tag{14}$$

Alas, this system has no solutions of the form $X = f(P)$, since the equations for the dynamics of the atmosphere and the ocean have no stationary solutions at $P = \text{const}$.

Even much simpler equations describing the motion of a spherical layer of viscous liquid on a rotating ball have a single stationary solution, namely, the "quasi-solid" behavior of the layer. But, as is known, it is unstable.

To apply the theory of singular perturbations to modeling a biospheric process one must also have at hand a time-averaged system, as in the case of Budyko's zero-dimensional climatic model. Unfortunately, for the time being, we cannot build time-averaged models of evolution of climatic characteristics based on relatively general (3- and 2-dimensional) hydro-thermodynamical equations.

In closing this section, I would like to emphasize two points peculiar to the problem of predictability of biospheric processes.

First, the equilibrium (or sustainability) of the "atmosphere + ocean + biota" system is evidently maintained by a very fine congruity between a multiplicity of factors. The result of their interaction is hard (perhaps, impossible) to predict more or less far into the future, for the inclusion of new components into a system's model can qualitatively shift the equilibrium condition.

For instance, in the zero-dimensional Budyko's model, equilibrium is always asymptotically stable. However, it can be easily demonstrated that once the glacial cover is accounted for (in the linear approximation), the single equilibrium becomes unstable. Further augmentation of the climatic model by inclusion of non-linear members and convective heat transfer (replacement of Budyko's zero-dimensional model with Sellers' one-dimensional one [Sellers, 1976]) maintains the unstable equilibrium at temperatures lower than $T = T_1$, but discloses the existence of stable equilibrium states at higher average temperatures. It is not inconceivable that in this stable state biospheric parameters will prove unsuitable for human life. The Earth will resemble Venus, though the oceans will remain. But virtually all carbon dioxide contained in the oceans will enter the atmosphere.

However, the inclusion of biotic elements in the model restores stability of the equilibrium state in the model of the "atmosphere + ocean + biota + glaciers + carbon dioxide" system at temperatures lower than $T = T_1$, while the sink whereby carbon dioxide leaves the ocean and enters the atmosphere has not become active.

As we have seen, the switch to three-dimensional models of the atmosphere, the ocean and the biota dynamics enabling more reliable predictions be made causes new difficulties: there are no stationary solutions at all, and we have to restrict ourselves to a numerical analysis of individual scenarios, and therefore drastically narrow down the possibility of understanding the most significant features of processes under review, that is, possible tendencies and trends. Under such circumstances, the scenarios invariably contain some uncertainty. Besides, the notion of sustainability requires special interpretation.

Under continuous external forcing, the processes of the atmosphere and ocean dynamics are, most likely, near-periodic functions of some kind (or, maybe, they are well-approximated by such functions over time intervals of interest). This line of inquiry into the problem of modeling of biospheric processes is undoubtedly very fruitful. At the same time, it is important to be able to extend the theory of singular perturbations to the area of near-periodic solutions of equations modelling quasi-equilibrium behavior. In a word, this line of investigation may provide a clue to predicting qualitative shifts in biospheric parameters occurring through anthropogenic pressure.

In summary, we can say that any categoric statement about the character of stability as well as predictions about the behavior of biospheric parameters should be made with much caution. Definitely, sustainable development of the biosphere is one of the most important problems of fundamental science.

3 Inaccuracy of Representation as a Mechanism of Stability

In most cases, the problem of prediction of natural processes is reduced to studying the patterns of changes in their characteristics under external forcing. The most interesting and practically important patterns are those concerned with stability, as we understand it, or, in other words, with the ability of a system under study to compensate for the effects of the environment that undermine its stability. These problems are often associated with feedback networks inherent in this system.

However, when it comes to nature, it is more appropriate to talk about a system's compensatory abilities than about feedback, for natural systems lack rationality.

Some particularities of "biospheric" models similar to Le Chatelier's principle were revealed in the previous section, where the matter was discussed in terms of the laws of conservation. Incidentally, the latter are not the only source of stability. Self-reproduction (reduplication) processes fill a highly important place both in the biota and in society. I am talking primarily of the reproduction of living matter, demographic processes included. But actually any organized structure is to an extent subject to reduplication. Reduplication never produces a precise copy, since the process inevitably involves fluctuations. Interestingly, it is due to these fluctuations that the organized structure of a self-reproducing ensemble is maintained, and therefore they can be looked upon as stabilizing factors of sorts. A simple example will give better insight into the matter.

Consider some set G of vectors s and define on it a non-negative scalar function $x(t, s)$ called medium density:

$$X = \{x(t, s)\} \,,$$

for example, biomass density. We call the set G the set of genetic characteristics of the medium x.

The condition of the medium, the function $x(t, s)$, varies with time under external forcing described by the vector $\xi(t)$ (with the understanding that it has the same dimensions as s) and by action of the internal mechanism of reduplication.

The change of the medium changing under the said forcing is given by a demographic equation of the form

$$\frac{\delta x(t, s)}{\delta t} = -f[\|\xi - s\|, x(t, s)] + \int_G \phi[s, c, x(t, c)]dc \,. \tag{15}$$

If, on the strength of Eq. (15) the quantity $x(t, s)$ proves negative for some constants c, we assume it to be zero. The integrand in the right-hand part of (15) contains the function $f[\|\xi - s\|, x]$ which increases monotonically with the norm of difference $\|\xi - s\|$ and vanishes at $x = 0$. This function gives the reduction of medium (biomass) density owing to the gap between the medium's genetic properties and external conditions. In terms of demography, this summand describes the "death rate". With regard to the addend, it describes self-reproduction (reduplication), that is, the "birth rate". The function ϕ is non-negative and reduces to zero at $x = 0$; it peaks at $s = c$, while at $s \neq c$ it describes "inaccuracy of reduplication" when an element of the medium possessing the property c reproduces elements with extra properties s.

Therefore, the equation (15) always admits the trivial solution $x = 0$. This is characteristic of demographic equations satisfying the Pasteur–Reddy principle: "The living can only come from the living".

If the reduplication were accurate, then the function ϕ would take on the properties of the δ-function, and the equation (15) would assume the form

$$\frac{\delta x}{\delta t} = -f[\|\xi - s\|, x(t, s)] + \phi[s, s, x(t, s)] \,. \tag{16}$$

In this case the medium would have stopped being a system and become a set of unconnected values $x(t, s)$, each varying independently with time.

A finite-dimensional analog of (15) is the following system of ordinary differential equations:

$$\frac{dy_k}{dt} = -f_k[\|\xi(t) - s_k\|, y_k] + \sum_c \phi[s_c, y_c] \,. \tag{17}$$

This system of equations is also bound to have a trivial solution

$$y_k(t) = 0 \,.$$

In its neighborhood, the system (17) takes the form:

$$\frac{dy_k}{dt} = -a_k \|\xi(t) + s_k\| y_k + \sum_c b_{ck} y_c , \tag{18}$$

where a_k and b_{ck} are some constants.

We now write out the finite-differential approximation of the system of equations (18):

$$y_k(t_{n+1}) = y_k(t_n) - a_k \|\xi(t_n) + s_k\| y_k(t_n) + \sum_c b_{ck} y_c(t_n) . \tag{19}$$

The model of a self-developing medium governed by (15), and, hence, by (18) and (19), has an intrinsic feedback mechanism, maintaining its wholeness (that is, sustainable development) and providing the adaptation necessary under changing environment parameters (the vector x). This can be verified using the finite-differential approximation (19) in a situation where $y_c(t_n) = 0$ for all c different from k. Then the system of equations (19) takes the form:

$$y_k(t_{n+1}) = y_k(t_n) - a_k \|\xi(t_n) - s_k\| y_k(t_n) + b_{kk} y_k(t_n) , \tag{20}$$

$$y_c(t_{n+1}) = b_{ck} y_k(t_n) , \quad c \neq k . \quad y_i \geq 0 ,$$

This situation is only possible if at instant t_{n-1} the condition of the environment (for example, temperature) changes in such a way that the norm of difference $\|\xi(t_{n-1}) - s_c\|$ grows to an extent at which all $y_c(t_c)$ vanish for all c different from k.

The latter does not imply that the element with the property s_c no longer exists in the system, for it will reappear in the next reduplication cycle. This is demonstrated by the formulas (20).

Moreover, the system's aggregate biomass $S = \sum_c y_c$ may exceed $S(t_n)$ in subsequent instants of time:

$$S(t_{n+1}) = \sum_c y_c(t_{n+1}) .$$

Hence, the system under review is controlled by a feedback mechanism maintaining, under certain conditions, its organized structure, that is, preserving the system's elements, its wholeness and enabling the growth of individual functionals.

The cited example shows that in order for a stabilizing feedback mechanism to be initiated in a system having the self-reproduction property, a "reduplication error" often suffices. The feedback is realized at the expense of changing ratios of elements possessing one or other property. Depending on the rate of change in the environment, the "reduplication error" must exceed one or other limit. The domain of admissible variations of the environmental characteristics (ξ) not disturbing a system's wholeness increases with growing measure of inaccuracy. However, the system's "effectiveness", when understood in terms of "biomass" production may decrease with growing measure of inaccuracy.

Notice that the ability to consume energy and matter from outside, that is, to effect metabolism as well as reduplication is inherent not only to living beings. M. Eigen was the first to model these processes for biological macromolecules [Eigen, 1971]. Hence, the occurrence of feedbacks capable of maintaining homeostasis, and thus different from Le Chatelier's principle does not apply strictly to living systems. The ability to maintain homeostasis inherent to any living being is not an immediate consequence of the laws of conservation, though, of course, it does not contradict them.

The structure of mechanisms responsible for the feedback still remains unclear. The cited example unveils to a certain degree its potential for its activity embedded by Nature in systems capable of reduplication.

The above reasoning and the example lead one to believe that the occurrence of other, more complexly organized feedback mechanisms can be explained in the context of empirical generalizations that form the basis for universal evolutionism [Moiseev, 1987, 1991].

The occurrence of the discussed mechanism securing sustainability is none other than a manifestation of "assembly algorithms" in action. These arise as a feature of a system independent of the properties of elements composing the system, but dependent on peculiarities of bonds that exist (or originate) between the elements. In the case discussed, such a mechanism is inherent in a system where the "assembly" is realized through the process of "inaccurate reduplication".

We now turn to the analysis of other characeristic properties of a self-reproducing system whose evolution is governed by equation (15). To study the behavior of its solutions in the neighborhood of the trivial solution $x = 0$, we restrict ourselves to its finite-dimensional analog, the system of equations (17). To this end, we use a standard logical technique: we discard nonlinear terms to analyze stability of stationary solutions. These solutions satisfy the equation:

$$a_k \| \xi - s_k \| y_k = \sum_j b_{jk} y_j . \tag{21}$$

The system of equations (21) has nontrivial solutions only when one of the quantities

$$\mu_k = a_k \| \xi - s_k \|$$

is equal to a positive eigenvalue of the matrix of (21): $\mu_k = \lambda_c$. We now arrange them according to absolute value: $\lambda_1, \lambda_2, \lambda_3, \ldots$

Bearing in mind that our analysis has an illustrative purpose, we assume the value ξ to be scalar (for example, it may be air temperature). We also assume the property s_k to be scalar (say, it may be a temperature most favorable for the existence of the population in question). We introduce the function $\Lambda_k(\xi)$ in the following way. For each value of i we work out

$$\mu_i(\xi) = |\xi - s_i|, \qquad \bar{\mu}(\xi) = \min \mu_i(\xi).$$

Now the function $\Lambda(\xi)$ can be defined as

$$\Lambda(\xi) = \min |\bar{\mu}_j(\xi) - \lambda_j|. \tag{22}$$

Nontrivial solutions of (21) are only possible for ξ "eigenvalues" corresponding to $\Lambda(\xi)$ zeroes, which we designate as $\xi_1, \xi_2, \xi_3, \ldots$.

In the neighborhood of values ξ_k corresponding to zero of the function $\Lambda(\xi)$, that is $\Lambda(\xi_k) = 0$, sustainability is no longer attainable, and nor is the ability to predict the behavior of the trajectory (17), because in the neighborhood of $\xi = \xi_k$, stationary solutions of the system of equations (18) undergo qualitative transformations. As a result, more than one possible evolution pattern occurs, and there appear structures separated from each other by unsurmountable barriers.

On the other hand, these stationary structures play the part of attractors. Once we know the attractors' topology and location of trajectories relative to them, we can make reliable predictions on the behavior of a certain trajectory.

The next stage of our study involves derivation of stationary solutions of the system (17), which would become trivial in the neighborhood of $\xi = \xi_k$.

Since the system (17) is nonlinear, we start with a search for a stationary solution in the neighborhood of a trivial one. To this end, it is only natural to use the Lyapunov–Schmidt method and to look for the stationary solution in the form of series arranged according to powers of the parameter ϵ, where

$$\epsilon = \mu_s - \lambda_n$$

and where the quantities s and n correspond to those value of ξ_k that allow $\Lambda = \Lambda_k$ to be realized. The solution sought will have the form:

$$y_i = \epsilon y_{i1} + \epsilon^2 y_{i2} + \ldots. \tag{23}$$

By substituting the series (23) into the equation (17), assuming $dy/dt = 0$, and developing the right-hand terms of (17) as a series in powers of ϵ, we obtain systems of linear equations useful for deriving terms in the series (23). As a first approximation, we get the system:

$$a_i \mu_i y_{i1} = \sum_j b_{ji} y_{ji}, \quad i = 1, 2, \ldots \tag{24}$$

where one of μ_i values is equal to λ_n, say, $\mu_s = \lambda_n$.

If the multiplicity of the proper number ξ_n is one, the solution is given by:

$$y_{i1} = C\bar{y}_{i1},$$

where C is an unspecified constant determined from a given solvability of the second-approximation equation, in which the rank of the expanded matrix of a second-approximation linear system equals that of the matrix (24). The subsequent order approximations are derived in a similar manner.

Next, we assume $\epsilon = \epsilon^*$, where ϵ^* is some arbitrarily taken, but small enough, number. It dictates the value $\xi^* = \xi_s + \epsilon^*$ and the corresponding value of the vector y with components $y_i = y_i(\xi)$. Using these values, we work through the corresponding Cauchy problem for the equation (17). Doing this enables a new nontrivial stationary solution $y^{(2)}(\xi)$ to be found.

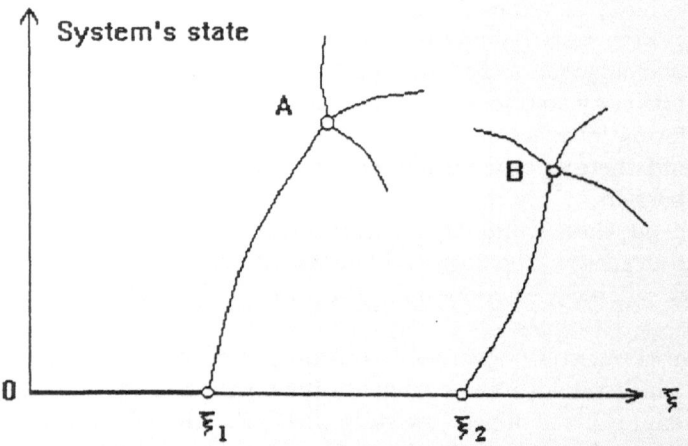

Fig. 2 As the external parameter ξ varies from 0 to the value ξ_1, at which the proper value $\Lambda(\xi)$ goes to zero, the system admits a trivial solution. Non-trivial stable conditions occur in the neighborhood of points $\xi_1, \xi_2, \ldots$ where $\Lambda(\xi)$ damps down to zero. At the far-from-equilibrium points A and B, proper values can become zero, too, thereby giving rise to new stable conditions in the neighborhood of A and B.

Linearizing (17) in the neighborhood of this solution we can repeat the procedure over and over again. As a result, we get the pattern of the arrangement of stationary solutions shown in Fig. 2.

Consider Fig. 2. As ξ varies on the interval $[0, \xi_1]$, where the function $\Lambda(\xi)$ has not reached the first zero $\Lambda(\xi_1)$ yet, the system has a trivial solution. It can be both stable and unstable. In the former case, the population will tend to die out, and there can be no question of sustainability, though the system will preserve wholeness in spite of progressive degradation. In the latter case, the system will retain sustainability and will evolve until it reaches some stable condition corresponding to the curve (ξ_1, A). In the neighborhood of the points A and B that are far from the equilibrium the function $\Lambda(\xi)$ can again damp down to zero, giving rise to new stable conditions in the neighborhood of A and B.

As we can see, conservation of wholeness and of sustainability are not the same notions characterising different features of evolving systems. Together they provide an insight into the process of self-development of a system capable of self-reproduction. The stationary conditions form a branching tree of sorts, or a skeleton acting as an attractor that guides development. In the course of self-organization, a system's condition comes close to one skeleton "branch" or another, depending on initial conditions. But by virtue of the probabilistic nature of the reduplication process, the system tends to retain its wholeness and to adapt the changing external conditions without breaking apart.

References

Budyko, M.I. (1980): *The heat balance of the globe.* Gidrometizdat Publishers: Leningrad (in Russian).

Eigen, M. (1971): *Self-organization of matter and the solution of biological macromolecules.* N.Y.

Kravtsov, Yu.A. (1989): "Randomness, determinateness and predictability". Sov. Phys. Uspekhi **32**(5), 424–449.

Kravtsov, Yu.A. (1993): "Fundamental and practical limits of predictability". In *The limits of predictability* (Ed. by Yu.A. Kravtsov). Springer, Heidelberg, Berlin, 173–203.

Moiseev, N.N., Aleksandrov, V.V., and Tarko, A.M. (1985): *The man and the biosphere*: Moscow.

Moiseev, N.N. (1987): *The algorithms of development.* Moscow.

Moiseev, N.N. (1991): "Universal evolutionism". Voprosi Filosofii, N3, 3–29.

Sellers, M.D. (1976): "A two-dimensional global climatic model". Month. Wea. Rev. **104**(3), 233–284.

Tikhonov, A.N. (1952): "Systems of differential equations containing a small parameter". Mathematical Works, 575–586 (in Russian).

Part 5

Analysis and Forecasting of Financial Data

The Application of Wave Form Dictionaries to Stock Market Index Data

James B. Ramsey and Zhifeng Zhang

Abstract

A matching pursuit algorithm is used to implement the application of wave form dictionaries to decompose the signal in the stock market (Standard and Poor's 500) index. A wave form dictionary is a class of transforms that generalizes both windowed Fourier transforms and wavelets. Each wave form is parametrized by location, frequency, and scale. Such transforms can analyze signals that have highly localized structures in either time or frequency space as well as broad band structures.

The Standard and Poor's 500 stock market index is found to be highly complex, but not a random walk. There are bursts of high energy that arise suddenly with very localized energy and die out equally quickly. In addition there is evidence of Dirac delta functions representing impulses, or shocks, to the system that seem to cluster more than would be expected under an hypothesis of random variation. It would appear that the energy of the system is largely internally generated, rather than the result of external forcing. Finally, there is apparently some evidence for a quasi-periodic occurrence of oscillations that are well localized in time, but that involve almost all frequencies.

1 Introduction

Two ideas are held to be tone when it comes to the behavior of stock market prices. The relative returns, or growth rates, are approximately a random walk; alternatively stated, prices are approximately a geometric Brownian motion [Fama, 1990; Fama and French, 1988a,b], and the series is nonstationary as indicated by the heteroskedastic behavior of the variances [Bollerslev et al., 1990; LeBaron, 1992; Poterba et al., 1986]. Consequently, to analyze such data requires techniques that allow for the presence of nonstationarity, do not make excessive demands on the data, and do not require knowledge of the detailed structure of the generating process which the economist does not have.

Traditional methods, such as correlation analysis, [Fama, 1990]; ARIMA and ARCH models, [Mankiw et al., 1991; Bollerslev et al., 1990], and spectral methods, [Granger and Morgenstern, 1963; Daniel and Torous, 1990], have corroborated in large part the basic assumptions made above, although there is evidence that the Brownian motion idea is only approximately right [Fama and French, 1988; LeBaron, 1990; Poterba and Summers, 1986]. Recently, ideas of "mean reversion" and "special day of the week" and "end of the month" effects have clarified the degree of approximation that is involved [Kim et al., 1991; Cutler et al., 1989; Schwert, 1989]. There is very little evidence for spectral power at almost any frequency, although there is a surprising degree of coherence between indices of monthly production and monthly values of the Standard and Poor's stock market index, [see for example, Ramsey and Thomson, 1994]. Finally, some recent work by [Ramsey and Zaslavsky, 1994] using wavelet analysis indicates that the stock market data are not random walks, but that the evidence for structure is at best weak; the data are complex for sure.

Recently, Stephen Mallat and colleagues have pioneered the estimation of dictionaries of wave forms [Mallat and Zhang, 1993]. This approach is a generalization of both Fourier analysis and of wavelet analysis [Daubechies, 1991]. As such the approach is particularly suited to the difficulties faced by an economist trying to understand the behavior of stock market prices. As will be explained in the next section, the benefit from the more general approach is that one can allow for all frequencies up to the Nyquist, including Dirac delta functions, and more importantly, the frequency evaluations can be localized in time and by scale of the windowing function.

This short chapter is in four sections. Section 2 will review the properties of dictionaries of wave forms, establish some notation, and indicate how to estimate the coefficients in the wave form approximation. The third section will report on the results of applying the analysis to the S and P 500 stock market index. The last section provides a summary of the results and explores the implications for future research.

2 Matching Pursuit and Dictionaries of Waveforms

Many signals are not stationary. Indeed, some signals, such as speech and perhaps financial data, are highly nonstationary and the nonstationarity often involves intermixing Dirac delta function impulses with discrete shifts in frequencies. Linear expansions in terms of a single basis, whether it is a Fourier, wavelet, or any other basis, are not flexible enough. The basis that might be suitable for one part of the data, may be most unsuitable for another part of the data. A Fourier basis provides a poor representation of signals that are tightly localized in time; a wavelet basis is not well adapted to represent functions whose Fourier representations are highly localized in frequency space. Each of these examples of an expansion basis for representing signals works well for the types of signals for which they were designed. But if one suspects either that the signal is highly nonstationary, or that a signal is a mixture of discrete and continuous changes, then more robust, but less specific, tools of analysis are needed. Flexible decompositions are important for representing signals that are characterized by localizations in time and frequency that vary widely over the whole signal. Impulses need to be decomposed over functions that are well concentrated in time; while spectral lines that are well localized in frequency are best represented by waveforms that have a narrow frequency support. Unfortunately, one cannot obtain high resolution in both time and frequency spaces at the same location in time-frequency space. Consequently, a local choice has to be made as to whether time or frequency localization will best represent the signal. This means that a flexible method is required.

The analytical approach used in this paper generalizes both windowed Fourier transforms and wavelets. The former approach enables one to estimate frequencies locally and to separate nearby frequencies. The latter approach concentrates on the effects of scaling of the data and enables one to estimate local time effects and to separate nearby impulses. Using wave form dictionaries, we will be able to handle both difficulties in one transformation.

2.1 Definition of Time-Frequency Atoms

A general family of time-frequency "atoms" can be generated by scaling, translating, and modulating a single window function $g(t) \in L^2(R)$. We suppose that the windowing function $g(t)$ is real and centered at 0. We also impose the conditions that $\|g\| = 1$, where $\| \cdot \|$ represents a suitable norm, that the integral of $g(t)$ is non-zero and that $g(0) \neq 0$. For any scale parameter $s > 0$, frequency modulation ξ, and translation u, we define the triplet $\gamma = (s, u, \xi)$ and define the "atom", $g_\gamma(t)$ by :

$$g_\gamma(t) = \frac{1}{\sqrt{s}} g\left(\frac{t-u}{s}\right) e^{i\xi t}. \tag{1}$$

The index γ is an element of the set $\Gamma = R^+ \times R^2$. The factor $1/\sqrt{s}$ normalizes the norm of $g_\gamma(t)$ to 1. The function $g_\gamma(t)$ is centered at the abscissa u and its energy is concentrated in a neighborhood of u, whose size is proportional to s. Its Fourier transform is centered at the frequency $\omega = \xi$ and has an energy concentrated in a neighborhood of ξ, whose size is proportional to $1/s$.

The dictionary of time-frequency atoms D is defined by $D = \{g_\gamma(t)\}_{\gamma \in \Gamma}$. The dictionary so defined is a highly redundant set of functions that includes window Fourier frames and wavelet frames [Daubechies, 1991]; redundancy means that the set Γ is not restricted to a basis set. When the signals include time-frequency structures of very different types, that is, when non-stationarity is an important and significant aspect of the signal, one cannot choose a priori a single frame that is well adapted to perform the expansion for all the constituent structures. Rather, for any given signal, we need to find a sequence of atoms from the dictionary that best match the signal structures in order to obtain a compact decomposition. In the next section we study such adaptive decompositions from redundant dictionaries.

2.2 Matching Pursuit

Let H represent a signal space. We define a dictionary as a family $D = \{g_\gamma(t)\}_{\gamma \in \Gamma}$ of vectors in H, such that $\|g_\gamma\| = 1$. We impose the condition that the linear expansion of vectors in D is dense in H. In general D is a redundant family of vectors that contains more vector elements than are required for a basis. The smallest complete dictionaries are bases. Consequently, when a dictionary is redundant a signal will not have a unique representation as a sum of vector elements, or atoms. Unlike the case of restricting attention to a basis for the space H, we have some degrees of freedom in choosing a signal's particular representation. This freedom allows us to choose a subset of the dictionary that is tailored to the signal in question and which will provide the most compact representation. We choose that subset of the dictionary for which the signal energy is concentrated in as few terms as possible. The chosen atoms highlight the dominant signal features as measured by the energy of the signal captured by the atoms.

Let D be a dictionary of vectors in H. An optimal approximation of $f \in H$ is the expansion:

$$\tilde{f} = \sum_{n=1}^{N} a_n g_{\gamma n},\tag{2}$$

where N is the number of terms in the expansion for a given degree of approximation; a_n and $g_{\gamma n} \in D$ are chosen in order to minimize $\|f - \tilde{f}\|$, where $\|\cdot\|$ represents a suitable norm.

Because of the impossibility of computing numerically an optimal solution, we utilize a "greedy algorithm" [Mallat and Zhang, 1993] that computes a useful sub-optimal approximation. Let $f \in H$. We want to compute a linear expansion of f over a set of vectors selected from D in such a way as to capture best the signal's inner structure. This is accomplished by successive approximations of f through orthogonal projections of the signal onto elements of D. Let $g_{\gamma 0} \in D$. The vector f can be decomposed into

$$f = \langle f, g_{\gamma 0} \rangle g_{\gamma 0} + Rf,\tag{3}$$

where Rf is the residual vector after approximating f in the direction of $g_{\gamma 0}$. Clearly $g_{\gamma 0}$ is orthogonal to Rf and is itself normalized to 1, so that:

$$\|f\|^2 = |\langle f, g_{\gamma 0} \rangle|^2 + \|Rf\|^2.\tag{4}$$

To minimize $\|Rf\|$, we must choose $g_{\gamma 0} \in D$ such that $|\langle f, g_{\gamma 0} \rangle|$ is maximum. In some cases, it is only possible to find a vector $g_{\gamma 0}$ that is almost the best in the sense that

$$|\langle f, g_{\gamma 0} \rangle| \geq \alpha \sup_{\gamma \in \Gamma} |\langle f, g_\gamma \rangle|,\tag{5}$$

where α is an optimality factor that satisfies $0 < \alpha \leq 1$.

A matching pursuit is an iterative algorithm that at successive stages decomposes the residue Rf from a prior projection by projecting that residue onto a vector of D, as was done for f. This procedure is repeated for each residue that is obtained from a prior projection. Let $R^0 f = f$. Suppose that the nth order residue $R^n f$, for some $n \geq 0$ has been computed. At the next stage we choose an element $g_{\gamma n} \in D$ which approximates the residue $R^n f$:

$$|\langle R^n f, g_{\gamma n} \rangle| \geq \alpha \sup_{\gamma \in \Gamma} |\langle R^n f, g_\gamma \rangle| . \tag{6}$$

The residue $R^n f$ is itself decomposed into

$$R^n f = \langle f, g_{\gamma n} \rangle g_{\gamma n} + R^{n+1} f, \tag{7}$$

which defines the residue to order $n + 1$. Since $R^{n+1} f$ is orthogonal to $g_{\gamma n}$

$$\| R^n f \|^2 = |\langle R^n f, g_{\gamma n} \rangle|^2 + \| R^{n+1} f \|^2 . \tag{8}$$

If this decomposition is continued up to order m, then f has been decomposed into the concatenated sum

$$f = \sum_{n=0}^{m-1} (R^n f - R^{n+1} f) + R^m f . \tag{9}$$

Substituting equation (6) into (8) yields

$$f = \sum_{n=0}^{m-1} \langle R^n f, g_{\gamma n} \rangle g_{\gamma n} + R^m f . \tag{10}$$

Similarly, $\| f \|^2$ is decomposed in terms of the concatenated sum

$$\| f \|^2 = \sum_{n=0}^{m-1} (\| R^n f \|^2 - \| R^{n+1} f \|^2) + \| R^m f \|^2 . \tag{11}$$

Equation (7) yields an energy conservation equation

$$\| f \|^2 = \sum_{n=0}^{m-1} \langle R^n f, g_{\gamma n} \rangle^2 + \| R^m f \|^2 . \tag{12}$$

Although the decomposition is nonlinear, we maintain an energy conservation as if it were a linear orthogonal decomposition. A major task is to understand the behavior of the residue $R^m f$ as m increases. By transposing a result proved by [Jones, 1987] for projection pursuit algorithms [Huber, 1985] one can prove that the matching pursuit algorithm as outlined above does converge, even in infinite dimensional spaces [Mallat and Zhang, 1993]. The following theorem is from [Mallat and Zhang, 1993].

Theorem 1 *Let $f \in H$. The residue defined by the induction equation (6) satisfies*

$$\lim_{m \to +\infty} \|R^m f\| = 0 \,. \tag{13}$$

Hence

$$f = \sum_{n=0}^{+\infty} \langle R^n f, g_{\gamma n} \rangle g_{\gamma n} \,, \tag{14}$$

and

$$\|f\|^2 = \sum_{n=0}^{+\infty} \langle R^n f, g_{\gamma n} \rangle^2 \,. \tag{15}$$

When H has finite dimension, $\|R^m f\|$ decays exponentially to zero.

2.3 Implementation of a Matching Pursuit Algorithm

When the dictionary is very redundant, the search for the vectors that best match the signal residues can be limited to a sub-dictionary $D_\alpha = \{g_\gamma\}_{\gamma \in \Gamma} \subset D$. We suppose that Γ_α is a finite index set included in Γ such that for any $f \in H$

$$\sup_{\gamma \in \Gamma_\alpha} |\langle f, g_\gamma \rangle| \alpha \geq \sup_{\gamma \in \Gamma} |\langle f, g_\gamma \rangle| \,. \tag{16}$$

Depending upon the chosen value of α and the degree of dictionary redundancy, the set Γ_α can be made much smaller than Γ. The matching pursuit is initialized by computing the inner product $(\langle f, g_\gamma \rangle)_{\gamma \in \Gamma_\alpha}$. The process continues by induction. Suppose that we have already computed $(\langle R^n f, g_\gamma \rangle)_{\gamma \in \Gamma_\alpha}$, for $n \geq 0$. We search in D_α for an element $g_{\tilde{\gamma} n}$ such that

$$|\langle R^n f, g_{\tilde{\gamma} n} \rangle| = \sup_{\gamma \in \Gamma_\alpha} |\langle R^n f, g_\gamma \rangle| \,. \tag{17}$$

In order to find a dictionary element that approximates f better than $g_{\tilde{\gamma} n}$, we search using a Newton method for an index γ_n within a neighborhood of $\tilde{\gamma}_n$ in Γ where $|\langle f, g_{\gamma n} \rangle|$ reaches a local maximum. The optimization produces the following inequality sequence:

$$|\langle R^n f, g_{\gamma n} \rangle| \geq |\langle R^n f, g_{\tilde{\gamma} n} \rangle| \geq \alpha \sup_{\gamma \in \Gamma} |\langle R^n f, g_\gamma \rangle| \,. \tag{18}$$

Once the vector $g_{\gamma n}$ is selected, we compute the inner product of the new residue $R^{n+1} f$ with any $g_\gamma \in D_\alpha$, with an updating formula derived from equation (6)

$$\langle R^{n+1} f, g_{\tilde{\gamma} n} \rangle = \langle R^n f, g_\gamma \rangle - \langle R^n f, g_{\gamma n} \rangle \langle g_{\gamma n}, g_\gamma \rangle \,. \tag{19}$$

Since both $\langle R^n f, g_\gamma \rangle$ and $\langle R^n f, g_{\gamma n} \rangle$ have been stored, this update requires us to compute only the expression $\langle g_{\gamma n}, g_\gamma \rangle$. Dictionaries are generally built so that this inner product is recovered with a small number of operations. Mallat and Zhang, 1993 describe how to compute efficiently the inner product of two discrete Gabor atoms in order (1) operations.

The number of times we sub-decompose the residues of a given signal f depends upon the desired precision ϵ. The number of iterations is the minimum p such that

$$\|R^p f\| = \left\| f - \sum_{n=0}^{p-1} \langle R^n f, g_{\gamma n} \rangle g_{\gamma n} \right\| \le \epsilon \|f\|. \tag{20}$$

The energy conservation equation (11) proves that this last equation is equivalent to

$$\|f\|^2 - \sum_{n=0}^{p-1} |\langle R^n f, g_{\gamma n} \rangle|^2 \le \epsilon^2 \|f\|. \tag{21}$$

Since we do not compute the residue $R^n f$ at each iteration we test the validity of (20) in order to stop the process.

2.4 Matching Pursuit with Time-Frequency Dictionaries

For dictionaries of time-frequency atoms, a matching pursuit yields an adaptive time-frequency transform. It decomposes any function $f(t) \in L^2(R)$ into a sum of complex time-frequency atoms that best match its residues. This section studies the properties of this particular matching pursuit decomposition. We derive a new type of time-frequency energy distribution by summing the Wigner distribution of each time-frequency atom.

Since a time-frequency atom dictionary is complete, Theorem 1 proves that a matching pursuit decomposes any function $f(t) \in L^2(R)$ into

$$f = \sum_{n=0}^{+\infty} \langle R^n f, g_{\gamma n} \rangle g_{\gamma n}, \tag{22}$$

where $\gamma_n = (s_n, u_n, \xi_n)$ and

$$g_{\gamma n}(t) = \frac{1}{\sqrt{s_n}} g\left(\frac{t - u_n}{s_n}\right) e^{i\xi_n t}. \tag{23}$$

The term $\langle R^n f, g_{\gamma n} \rangle$ in equation (22) is the same term as "a_n" in equation (2).

These atoms are chosen in sequence to approximate the sequence of residues of f. For signals of size N, we can discretize γ and facilitate the computations in the following manner. We redefine γ by

$$\gamma = (2^j, 2^{j-1}p, k\pi 2^{-j}) \tag{24}$$

where $s = 2^j, u` = 2^{j-1}p, \xi = k\pi 2^{-j}$, and (j, p, k) are integers; $0 < j < \log_2 N, 0 \leq p < N2^{-j-1}, 0 \leq k < 2^{j+1}$. Using these definitions, equation (22) for the nth atom becomes:

$$\boldsymbol{g}_{\gamma_n}(t) = \frac{1}{\sqrt{2^{j_n}}} g\left(\frac{t - 2^{j-1}p_n}{2^{j_n}}\right) e^{ik_n\pi 2^{-j_n}t}, \tag{25}$$

where j_n, p_n and ξ_n are the values taken by the triplet (j, p, ξ) for the nth atom.

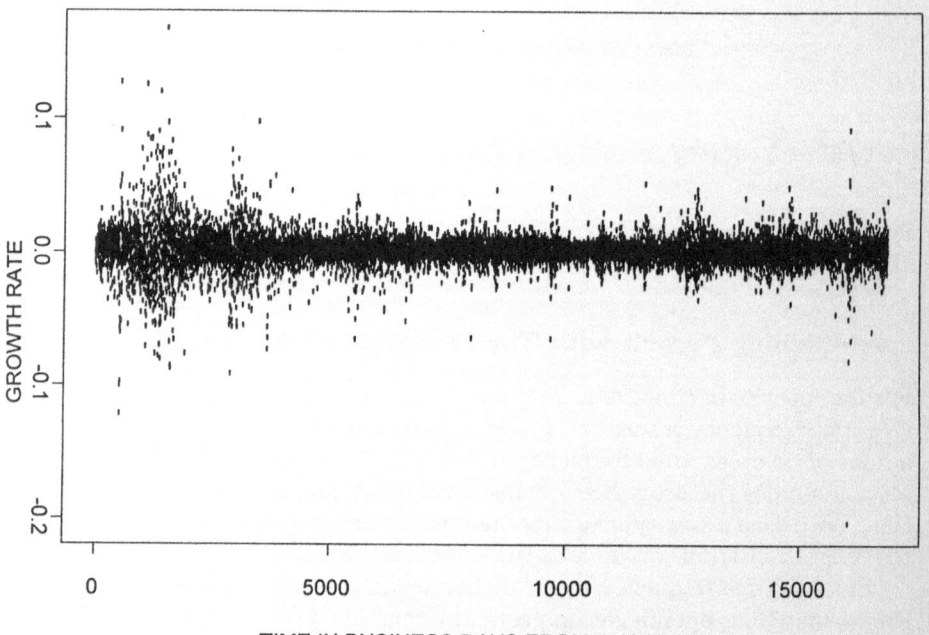

Fig. 1 Standard & Poor's 500 index: growth rates.

From the decomposition of any $\boldsymbol{f}(t)$ within a time-frequency dictionary, we derive a new time-frequency energy distribution, by adding the Wigner distribution for each selected atom. The cross Wigner distribution for two functions $\boldsymbol{f}(t)$ and $\boldsymbol{h}(t)$ is defined by:

$$W[\boldsymbol{f}, \boldsymbol{h}](t, \omega) = \frac{1}{2\pi} \int_{-\infty}^{+\infty} f(t + \tau/2)\bar{h}(t - \tau/2)e^{-i\omega\tau}d\tau. \tag{26}$$

The Wigner distribution of $\boldsymbol{f}(t)$ is $W\boldsymbol{f}(t, \omega) = W[\boldsymbol{f}, \boldsymbol{f}](t, \omega)$. Since the Wigner distribution is quadratic, we derive from the atomic decomposition (21) of $\boldsymbol{f}(t)$ that

$$
W\boldsymbol{f}(t,\omega) = \sum_{n=0}^{+\infty} |\langle \boldsymbol{R}^n \boldsymbol{f}, \boldsymbol{g}_{\gamma n} \rangle|^2 W\boldsymbol{g}_{\gamma n}(t,\omega)
$$

$$
+ \sum_{n=0}^{+\infty} \sum_{m=0, m\neq n}^{+\infty} \langle \boldsymbol{R}^n \boldsymbol{f}, \boldsymbol{g}_{\gamma n} \rangle \overline{\langle \boldsymbol{R}^m \boldsymbol{f}, \boldsymbol{g}_{\gamma m} \rangle} W[\boldsymbol{g}_{\gamma n}, \boldsymbol{g}_{\gamma m}](t,\omega) \,.
$$

(27)

The double sum corresponds to the cross terms of the Wigner distribution. It regroups the terms that one usually tries to remove in order to obtain a clear picture of the energy distribution of $\boldsymbol{f}(t)$ in the time-frequency plane. We thus only keep the first sum and define

$$
E\boldsymbol{f}(t,\omega) = \sum_{n=0}^{+\infty} |\langle \boldsymbol{R}^n \boldsymbol{f}, \boldsymbol{g}_{\gamma n} \rangle|^2 W\boldsymbol{g}_{\gamma n}(t,\omega) \,.
$$

(28)

A similar decomposition algorithm over time-frequency atoms was derived independently by [Qian and Chen, 1994], in order to define this energy distribution in the time-frequency plane. From the well known dilation and translation properties of the Wigner distribution and the expression (22) of a time-frequency atom, we derive that for $\gamma = (s, \ u, \ \xi)$

$$
W\boldsymbol{g}_{\gamma}(t,\omega) = W\boldsymbol{g}\left(\frac{t-u}{s}, s(\omega - \xi)\right) \,,
$$

(29)

and hence

$$
E\boldsymbol{f}(t,\omega) = \sum_{n=0}^{+\infty} |\langle \boldsymbol{R}^n \boldsymbol{f}, \boldsymbol{g}_{\gamma n} \rangle|^2 W\boldsymbol{g}\left(\frac{t-u_n}{s_n}, s_n(\omega - \xi_n)\right) \,.
$$

(30)

The Wigner distribution also satisfies:

$$
\int_{-\infty}^{+\infty} \int_{-\infty}^{+\infty} W\boldsymbol{g}(t,\omega)dtd\omega = \|\boldsymbol{g}\|^2 = 1 \,,
$$

(31)

so that the energy conservation equation (14) implies

$$
\int_{-\infty}^{+\infty} \int_{-\infty}^{+\infty} E\boldsymbol{f}(t,\omega)dtd\omega = \|\boldsymbol{f}\|^2 \,.
$$

(32)

We can interpret $E\boldsymbol{f}(t,\omega)$ as an energy density of $\boldsymbol{f}$ in the time-frequency plane (t,ω). Unlike the Cohen class distributions, it does not include cross product terms. It also remains positive if $W\boldsymbol{g}(t,\omega)$ is positive, which is the case when $\boldsymbol{g}(t)$ is Gaussian. On the other hand, the energy density $E\boldsymbol{f}(t,\omega)$ does not satisfy marginal properties, as opposed to certain Cohen class distributions [Cohen, 1989]. The importance of these marginal properties for signal processing is however not clear. If $\boldsymbol{g}(t)$ is the Gaussian window

$$
\boldsymbol{g}(t) = 2^{1/4} e^{-\pi t^2} \,,
$$

(33)

then

$$Wg(t,\omega) = 2e^{-2\pi\left(t^2 + \left(\frac{\omega}{2\pi}\right)^2\right)}.$$ (34)

The time-frequency atoms $g_\gamma(t)$ are then called Gabor functions. The time-frequency energy distribution $Ef(t,\omega)$ is a sum of Gaussian pulses whose locations and variances along the time and frequency axes depend upon the parameters (s_n, u_n, ξ_n).

3 The Empirical Results

The data that were used are the relative first differences, or growth rates, of an index of the end of day prices on five hundred stocks that are traded on the New York Stock Exchange; this index is known as the Standard and Poor's 500 stock index. There are 16384, or 2^{14}, observations. The historical period covered is January 3, 1928 to November 18, 1988. A time series plot of the growth rates is shown in Fig. 1. The data are characterized by a dense core of rapid oscillations interspersed with episodic bursts of oscillations of very large amplitude in some cases and many bursts of substantial variance relative to the ambient variation.

Given the discussion in the previous section, the behavior represented in Fig. 1 indicates that analysis by means of dictionaries of wave forms might well prove to be a useful exploratory tool. The major reason is that prior analysis of these data indicate that Dirac delta functions might play a major role and that, while oscillations at specific frequencies over the entire time domain are unlikely, specific frequencies occurring in short bursts, that is, "chirps", might well be observable. The result of running the matching pursuit algorithm on the dictionary of wave forms is illustrated in Figs. 2 and 3. In both figures, "time" runs from left to right along the ordinate, and frequency in cycles per month is represented on the abscissa from zero to one half with zero frequency at the bottom.

Figure 2 shows the Wigner distribution estimated over the 2048 days from November 1934 to September 1941. The high energy line in the middle of the graph represents the 1939 stock market collapse. Figure 3 represents 2048 observations from November 1980 to June 1990 and the high energy line just to the right of the center represents the October 1987 crash.

Both figures are representative of all the figures that were created. The bulk of the total energy represented in the graphs is contained in either Dirac delta functions, or in very narrow bands of energy along the time axis. There is some evidence of chirps in both graphs that are concentrated about the isolated high energy lines; there are also some very low intensity chirps that are distributed at random.

The major axis for the chirps is predominantly parallel to the frequency axis; that is, there are bursts of groups of frequencies that are well localized

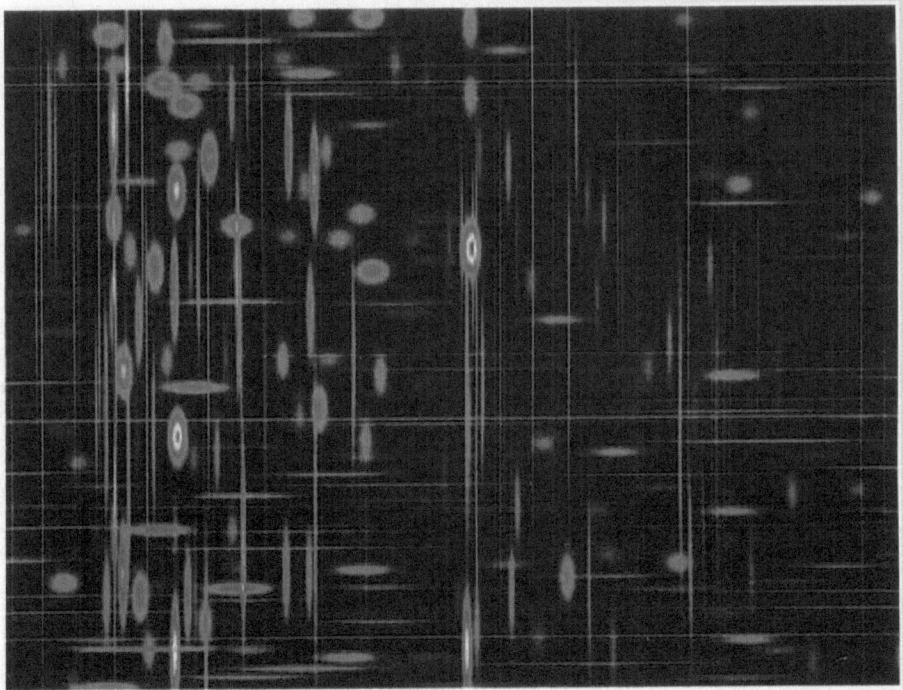

Fig. 2 Time-frequency Wigner distributions for Standard & Poor's 500 index I;
Axes: x: Time in working days from Dec. 03 '34 to Sept. 22 '41; y: Frequency:
range from 0 to 1/2.

in time. This observation is less true in the 1980 to 1990 period. In the later
period there is more evidence of frequencies holding for some months at a
time, dying out, then recurring, but not at regular intervals.

The localized bursts of high intensity energy have no advance build up in
intensity and die out equally abruptly. When eruptions do occur, the energy is
distributed over most of the frequency range, although relatively less energy
is observed at very low and very high frequencies.

Except for the areas of high energy bursts, the wave form distribution is
similar to that of noise. The distribution of relative intensities, as indicated
by the variation in shading, for the Standard and Poor's stock index plots
indicate that these are not random data, even if they seem to have a very
complicated structure. Random data tend to have a much more uniform
distribution of intensities, whereas the stock data have periods of relative
quiescence interspersed by periods of high intensity oscillations. In addition,
the more intense activity areas cluster far more with the stock data than is
true for random data.

Figure 4 portrays in descending value the magnitudes of the weight functions "a_n" that are defined in equations (2) and (22). Two sequences of

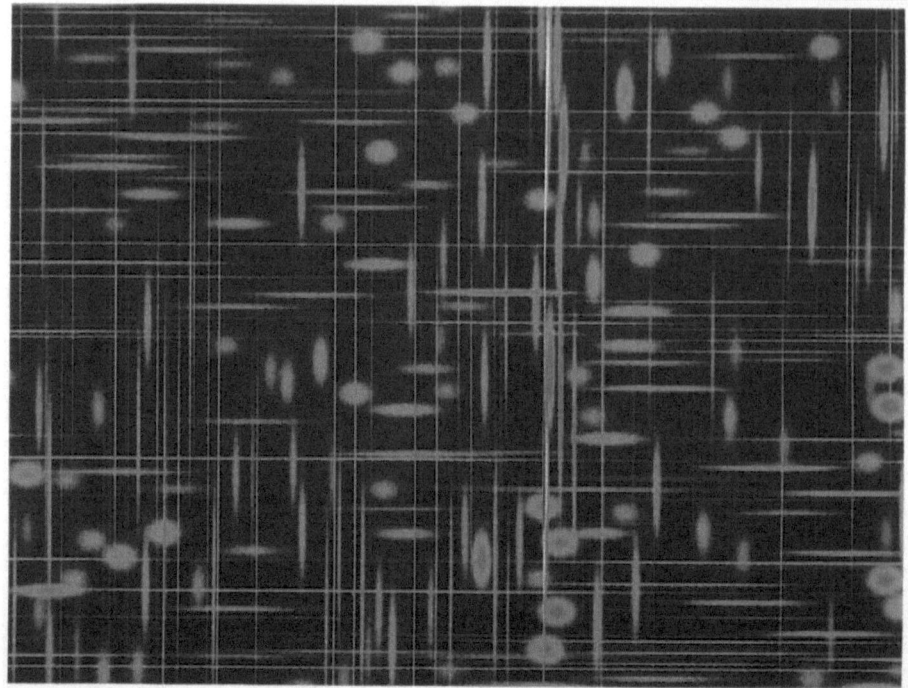

Fig. 3 Time-frequency Wigner distributions for Standard & Poor's 500 index II; Axes: x – time in working days from Dec. 01 '80 to Nov. 18 '88; y — frequency: range from 0 to 1/2.

weights are shown, one for the S & P 500 index and one for the same number of observations on a normal distribution; in this latter case, the number of waveforms that were calculated was three thousand, the same as for the S & P data. Consequently, the two graphs are, except for scale, identical in construction.

A data series with simple structure can be easily represented by a relatively small number of wave forms, so that the shape of the plot of weights is one of rapidly declining levels. In comparing the shape of the weight curves for the normal noise driven results and for the S & P results, we see that despite the lack of evidence for any structure in the data so far, there is relatively more rapid decline in the values of the weights for the S & P index. The S & P is well approximated by the first thousand wave forms, whereas the noise results require nearly three thousand to achieve the same degree of approximation.

We have pointed out that the majority of the power in the wave forms is in time localized bursts or even in Dirac delta functions. Consequently, an immediate question is whether there is information in the distribution over time of the Dirac delta function wave forms. Figure 5 plots the occurrence of

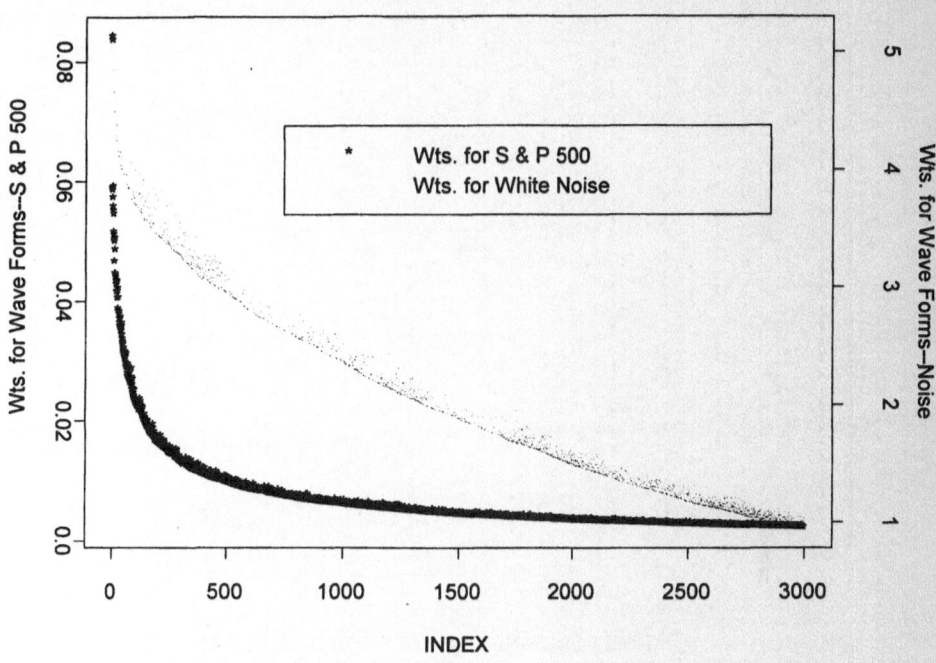

Fig. 4 Comparison of wave form weights, a_n.

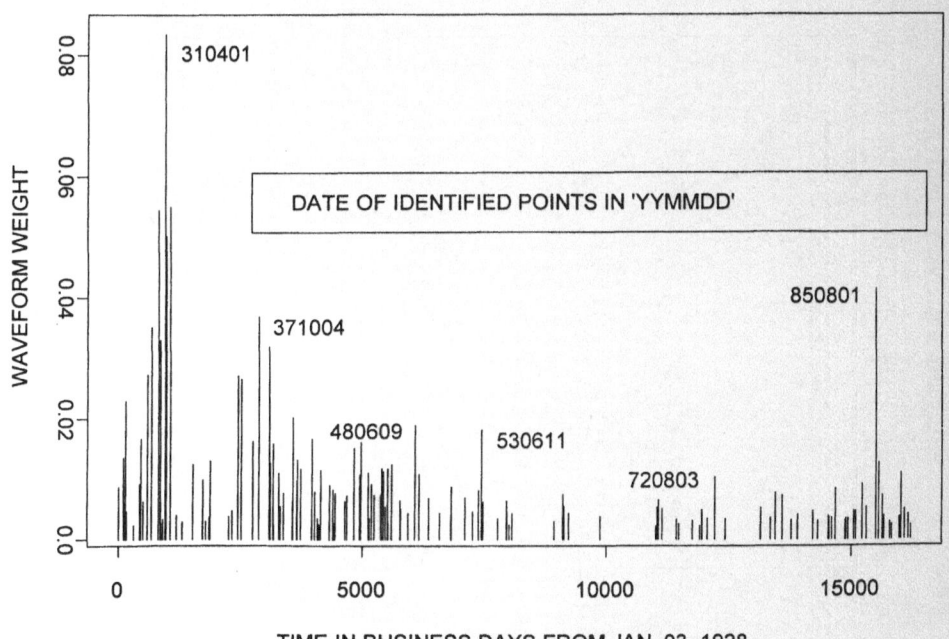

Fig. 5 Weights for Dirac delta function waveforms.

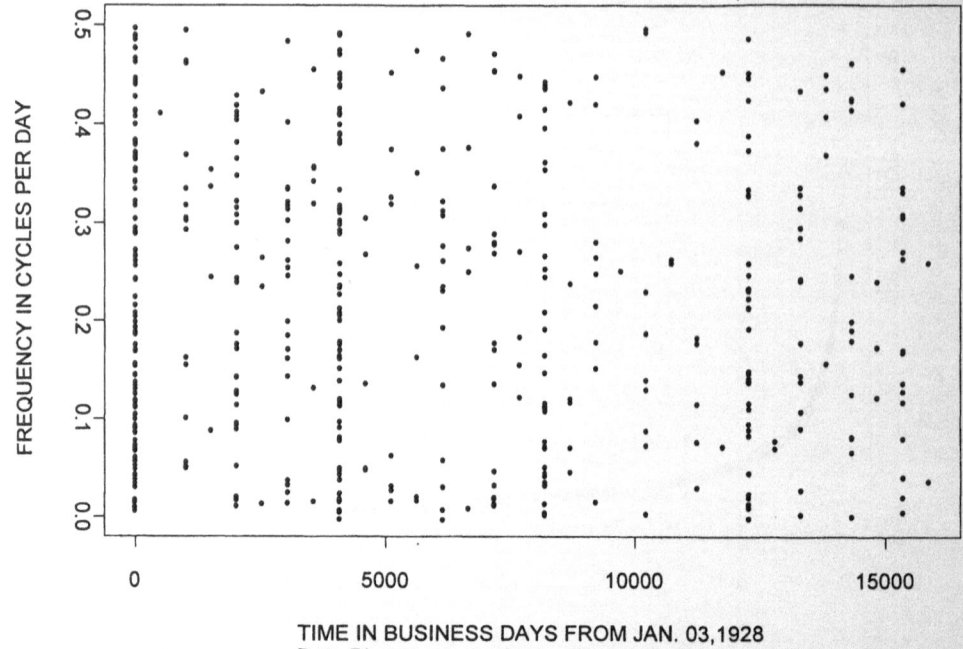

TIME IN BUSINESS DAYS FROM JAN. 03,1928
Data Plotted only for Scales Greater than 11

Fig. 6 Plot of frequencies by time period: Standard and Poor's index.

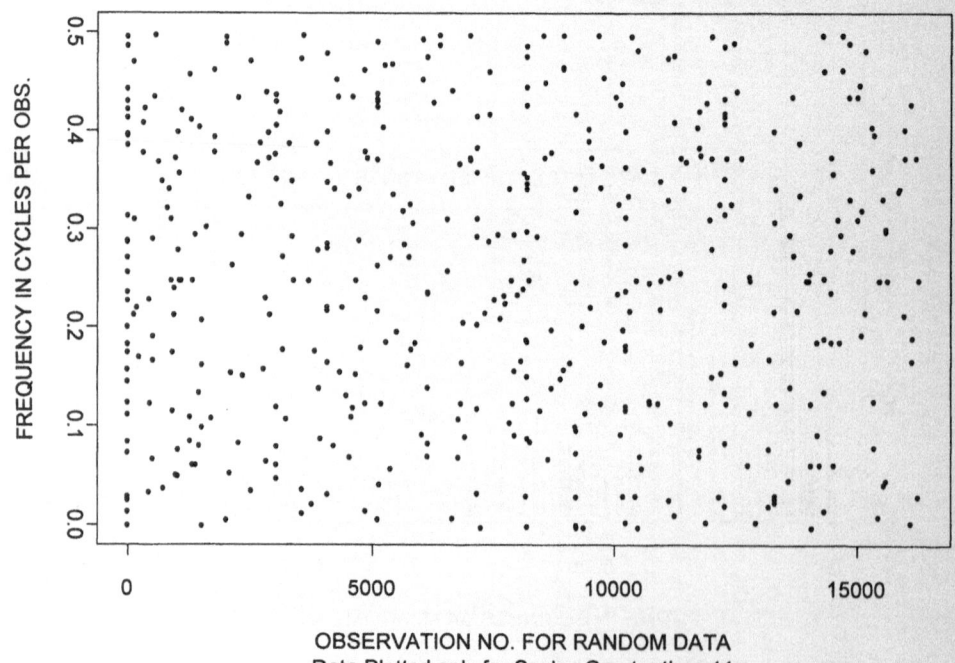

OBSERVATION NO. FOR RANDOM DATA
Data Plotted only for Scales Greater than 11

Fig. 7 Plot of frequencies by time period: random Gaussian data.

the Dirac delta functions over time. While there is no obvious regularity in the distribution of the delta functions, there is more clustering than would be obtained with noisy data. One interesting sidelight is that neither the October 1929, nor the October 1987, stock market crashes show up in the plot as highly significant delta functions, even though both periods are the centers of very narrow regions in the time domain with high bursts of energy. This indicates that the major stock market crashes and even the lesser ones, while of short duration, are not the output of isolated impulses.

From the discussion in the previous section, the reader will recall that the decomposition of the data by frequency components is best done at the higher scales, that is, for larger values for s_n; the frequencies are resolved more precisely than at the lower scales. Compare Figs. 6 and 7; the former shows the Wigner distribution of frequencies by position when the octave scales, "j" in equation (24), are restricted to the range of 12 to 14, 14 being the maximum level. The latter figure shows the exact same type of graph for random (normal) data. The greater degree of regularity in the Standard and Poor's stock market data is readily apparent. What may not be so clear visually is that many of the plotted points for the Standard and Poor's data seem to be quasi-periodic. While there is definitely greater regularity in the stock market data, we are not yet prepared to claim that the observed quasi-periodicity is a genuine phenomenon and not merely an artifact of our processing that we have not yet discovered. However, if these results are at least in part true, the implication is that quasi-periodically there are bursts of energy at almost all frequencies. Such results, coupled with no strong evidence of power at specific frequencies in between the bursts, would produce spectra that would be difficult to distinguish from a noise spectrum by conventional methods.

4 Summary and Conclusions

The results in this analysis have confirmed much of the current thinking about stock market data; they are not random walks, but the structure of the data is very complex. The analysis cited above has been able to clarify these vague notions and to refine some of the aspects of the complexity.

Our first piece of evidence that the data are not random is that the number of structures needed to provide a given level of decomposition is far fewer than is needed for random data. This result is even stronger when coupled with the recognition that the stock market data are best characterized by periods of relative quiet interspersed by periods of highly time localized intense activity. These periods of intense energy occur without any apparent warning, build in intensity very rapidly, and die out equally rapidly. Nonetheless, these bursts are not the result of isolated impulses that would be represented by Dirac delta functions. There are numerous cases of isolated impulses, but the occurrences of these are separate in general from the intense bursts. The occurrence of the Dirac delta functions represent impulses, or "shocks" to the system from external sources, whereas the energy bursts that are so prominent in the data seem to be internal to the system. The bursts contain high energy at most frequencies, relatively less at very low and very high frequencies.

There is also some tentative evidence that there are occurrences of activity across a range of frequencies at quasi-periodic intervals.

This research has posed more questions than it has answered. The most important finding that will stimulate much interest is that the occurrence of sudden bursts of energy at all frequencies perhaps should not be associated with isolated impulses, or shocks. While a speculative conclusion, it is reasonable to infer from these results that the bursts of intense activity are not the result of isolated, unanticipated, external shocks, but more likely the result of the operation of a dynamical system with some form of intermittency in the dynamics.

Acknowledgments

The financial and institutional support of the C.V. Center for Applied Economics is gratfully acknowledged.

References

Bollerslev, T., Chou, R. Y., Jayaraman, N., Kroner, K. F. (1990): "ARCH Modeling in Finance: A Review of the Theory and Empirical Evidence". Journal of Econometrics **52**(1), 5–60.

Cohen, L. (1989): "Time-frequency distributions: a review". Proceedings of the IEEE **77**(7), 941–979.

Cutler, D., Poterba, J.M., and Summers, L.H. (Spring 1989): "What Moves Stock Prices ?" Journal of Portfolio Management, 4–12.

Daniel, K. and Torous, W. (1990): "Common Stock Returns and the Business Cycle". UCLA Working Paper, Los Angeles, CA.

Daubechies, I. (1991): "Ten Lectures on Wavelets". CBMS-NSF Series in Appl. Math., SIAM.

Davis, G., Mallet, S., and Avenaleda, M. (1994): "Chaotic Adaptive Time-Frequency Decompositions". Technical Report, Computer Science, NYU.

Fama, E.F. and French, K.R. (1988): "Permanent and Temporary Components of Stock Prices". Journal of Political Economy **96**(2), 246–273.

Fama, E.F. (1990): "Stock Returns, Expected Returns, and Real Activity". The Journal of Finance **45**(4), 1089–1108.

Fama, E.F. and French, K.R, (1988): "Dividend Yields and Expected Stock Returns". Journal of Financial Economics **22**, 3–25.

Granger, C.W.J. and Morgenstern, O. (1963): "Spectral Analysis of New York Stock Market prices". Kyklos **16**, 1–27.

Huber, P.J. (1985): "Projection Pursuit". The Annals of Statistics, **15**(2), 435–475.

Jones, L.K. (1987): "On a conjecture of Huber concerning the convergence of projection pursuit regression". The Annals of Statistics **15**(2), 880–882.

Kim, M.J. and Nelson, C.H. (1991): "Mean Reversion in Stock Prices? A Reappraisal of the Empirical Evidence". Review of Economic Studies **58**, 515–528.

LeBaron, B. (1992): "Nonlinear Forecasts for the S & P Stock Index". In Martin Casdagli and Stephen Eubank, Eds., *Nonlinear Modeling and Forecasting, Diagnostic Testing for Nonlinearity, Chaos, and General Dependence in Time Series Data*, Redwood City, Ca.: Addison-Wesley, 381–394.

Mallet S. and Zhang Z. (1993): "Matching Pursuit with Time-Frequency Dictionaries". IEEE Trans. on Signal Processing, December.

Mankiw, N., Gregory, Romer D., and Shapiro M.D. (1991): "Stock Market Forecastability and Volatility: A Statistical Appraisal". Review of Economic Studies **58**, 455–577.

Poterba, J.M. and Lawrence, S.H. (1986): "The Persistence of Volatility and Stock Market Fluctuations". The American Economic Review **76**(5), 1142–1151.

Ramsey, J.B. and Thomson, D.J. (1994): "A Reanalysis of the Spectral Properties of Some Economic Time Series". Dept. of Economics, Working Paper Series, New York Univ., New York.

Ramsey, J.B. and Zaslavsky, G. (1994): "A Wavelet Analysis of U.S. Stock Price Behavior". Dept. of Economics Working Paper Series, New York Univ., New York.

Qian, S. and Chen, D. (1994): "Signal Representation via Adaptive Normalized Gaussian Functions". IEEE Trans. on Signal Processing **36**(1).

Part 6

Socio-Political
and Global Problems

Messy Futures and Global Brains

Gottfried Mayer-Kress

Abstract

The recent history after World War II was characterized by a relatively simple partition of the world in to basically two domains of superpower interests. Security issues could be discussed and analyzed in a global framework of two strategic players. There were clear goals and roles for the players. Today with the role of strategic nuclear weapons greatly reduced, we have regional crises which have some similarities with pre-World-War I situations with one major difference: Today's world is much more connected, especially with regard to information: On a large scale we are able to get direct first hand information from crisis areas and — for example, through computer networks — can directly participate in the discussion. That makes the future from a traditional control point of view *messy* and on a global scale more complex and less predictable. For that reason we think that the conditions for the emergence of a *Global Brain* will become a practical reality for global modeling and simulation in the very near future. We also discuss some of the potential future applications.

1 Introduction

In the framework of complex dynamical systems, we can view the world at the beginning of this century as loosely connected, with clusters of highly complex trouble spots and the number of people with a global perspective and fast connections was very small. Today we have multiple systems of global information exchange together with impressive data and information systems accessible for a rapidly growing number of people. We claim that the global computer network[13] will have a much more dramatic impact on global issues than other networks of global communication systems like television or telephone. In the first case we obtain many features that we can associate with the development of biological brains, whereas in the two latter cases many essential features are missing. The most relevant one is the creation of the analog of *cell assemblies*. We know from the theory of complex neural systems (both artificial and natural) that the connectivity between the individual information processing units is essential for solving large, global tasks. For that reason we think that the conditions for the emergence of a *Global*

---

[13]The Internet is currently growing at a rate of some 20% per month.

Brain will become a reality in the near future. In this paper we want to discuss the current status of three ingredients of a new modeling approach based on the *Global Brain* paradigm: distributed, associative information servers; global computer communication systems and their role as empirical research tool and finally distributed simulation servers which will develop into a complex, adaptive, and evolving global model of our world. It has become clear that our individual brain is not the stimulus/response automaton as which it was frequently seen. The adequate response to complex external stimuli would be too slow to be effective. Therefore there is a continuous unconscious modeling process going on in different levels of brain activity.[14] If the modeling of an input stream is successful, we don't have to pay too much attention to it: we can anticipate what will happen next. This phenomenon at a very low level is known as *habituation*. Noteworthy new information is associated with a *P300-alert*. The same mechanism seems to be at work at the perception of music and for the distinction between interesting and boring pieces of music [see e.g. Mayer-Kress, Bargar and Choi, 1992; Mayer-Kress et al., 1993; Mayer-Kress, Choi and Bargar, 1993; Mayer-Kress, 1994a and references therein]. If we want to succeed in the transition to a sustainable self-organized management of this planet, we have to make the transition from reactive *fixing of problems* to an active approach of modeling and control. In such a system the aspect of identification of situations with severely limited predictability is almost as important as the predictions themselves. Recognizing that a complete solution to all problems will not be feasible for the foreseeable future, we take a pragmatic perspective. Under the assumption that we have access to a well developed, global computer network we identify the following steps: (i) define targets for the solution of important problem areas (population, CO_2 level, violation of human rights, etc.) and assign a relevance weight to each of the problem areas; (ii) acquire qualitative information on the current status of the problem; (iii) define sub-areas where a quantitative approach appears to be promising; for those areas (iv) obtain current quantitative data and identify models that deal with the solution of any of the sub-problems; (v) create a conceptual model of the integrated system; (vi) link data and simulation models to an interdependent, distributed network; (vii) perform simulation, sensitivity analysis, and (viii) compare the results with the updated information from (ii) and (iii) and evaluate them with respect to the targets specified in (i). From the study of nonlinear and chaotic systems we know that only short-term predictions are possible if the system is complex and exhibits chaos. Therefore, a typical time scale of five years between formulation and verification of a global model appears to be too long in a world where time scales of, e.g., regional conflicts as in eastern Europe are significantly shorter than one year. Future models will have to be object-oriented with links to other models and information systems and they

---

[14]In a broad sense, that might be related to the important function of dreaming in biological brains.

will have to be adaptive to changing basic conditions. In this contribution we want to mainly focus on items (ii), (iv) and (v) in the above list.

2 What Is Chaos?

When we look at the changing world that we are living in, we can categorize the types of changes into a few fundamental categories: growth and recession, stagnation, cyclic behavior and unpredictable, erratic fluctuations. All of these phenomena can be described with very well developed linear mathematical tools. Here *linear* means that the result of an action is always proportional to its cause: if we double our effort, the outcome will also double. However, as Stan Ulam had pointed out, most of nature is non-linear in the same sense as "most of zoology is non-elephant zoology". The fact that most of traditional science is focusing on linear systems can be compared to the story of the person who looks for the lost car keys under a street lamp because it is too dark to see anything at the place where the keys were lost. Only recently do we have access to methods and computer power to make significant progress in the field of nonlinear systems and understand, for example, seemingly simple things like dripping faucets. One whole class of phenomena which does not exist within the framework of linear theory has become known under the buzz-word of *chaos*. The modern notion of chaos describes irregular and highly complex structures in time and in space that follow deterministic laws and equations. This is in contrast to the structureless chaos of traditional equilibrium thermodynamics. The basic example system that might be helpful for visualization, is a fluid on a stove, the level of stress is given by the rate at which the fluid is heated. We can see how close to equilibrium there exists no spatial structure, the dynamics of the individual subsystem is random and without spatial or temporal coherence. Beyond a given threshold of external stress, the system starts to self-organize and form regular spatial patterns (rolls, hexagons) which create coherent behavior of the subsystems ("order parameters enslave subsystems"). The order parameters themselves do not evolve in time. Under increasing stress the order parameters themselves begin to oscillate in an organized manner: we have coherent and ordered dynamics of the subsystems. Further increase of the external stress leads to bifurcations to more complicated temporal behavior, but the system as such is still acting coherently. This continues until the system shows temporal deterministic chaos. The dynamics is now predictable only for a finite time. This predictability time depends on the degree of chaos present in the system. It will decrease as the system becomes more chaotic. The spatial coherence of the system will be destroyed and independent subsystems will emerge which will interact and create temporary coherent structures. In a fluid we have turbulent cascades where vortices are created that will decay into smaller and smaller vortices. Analog situations in societies can be currently studied in the former USSR and eastern Europe. James Marti, 1991 speculates:

"Chaos might be the new world order". At the limit of extremely high stress we are back to an irregular Tohu-wa-Bohu-type of chaos where each of the subsystems can be described as random and incoherent components without stable, coherent structures. It has some similarities to the anarchy with which we started close to thermal equilibrium. Thus the notion of "chaos" covers the range from completely coherent, slightly unpredictable, strongly confined, small scale motion to highly unpredictable, spatially incoherent motion of individual subsystems. A scheme graphics of this organizational structure is displayed in Fig. 1.

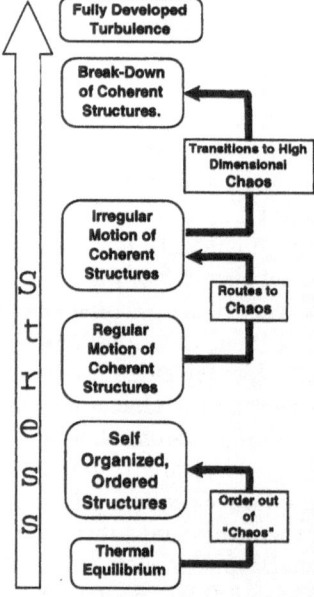

Fig. 1 Organization of chaos and turbulence.

There is frequent confusion between chaos and randomness. There are some similarities in the nature of chaotic and random systems, but there are also some fundamental differences. Some of them are listed in Table 1.

The game of roulette is an interesting example that might illustrate the distinction between random and chaotic systems: If we study the statistics of the outcome of repeated games, then we can see that the sequence of numbers is completely random. That led Einstein to remark: "The only way to win money in roulette is to steal it from the bank." On the other hand we know the mechanics of the ball and the wheel very well and if we could somehow measure the initial conditions for the ball/wheel system, we might be able to make a short term prediction of the outcome. Exactly this has been done by a group of Santa Cruz students who called themselves "Eudemonic En-

Table 1. Differences between order, chaos, and randomness. Planets used to be representations of a divine order. Chaotic signals can show spectra in the full range from pure tones to very noisy. The dimension of a dynamical system indicates the number of independent variables. An attractor determines the geometrical structure towards which a system will evolve

System	ORDER	CHAOS	RANDOMNESS
Paradigmatic example	Clocks, planets	Clouds, weather	Snow on TV screen
Predictability	Very high	Finite, short term	None→simple laws
Effect of small errors	Very high	Explosive	Nothing *but* errors
Spectrum	Pure	Yes!	Noisy, broad
Dimension	Finite	Low	Infinite
Control	Easy	Tricky, very effective	Poor
Attractor	Point, cycle, torus	Strange, fractal	No!

terprises". Their story of how they used chaos theory to conquer the casinos of Las Vegas and Atlantic City is described in [Bass, 1985].

3 Correlations Between Variables

Many problems connected with the analysis of complex, political issues are related to the questions of what factors are causing a specific desirable or undesirable development. One example might be to ask how environmental factors are affecting security issues. A traditional way of analysing this problem is to compare histories of each of the parameters and perform statistical analysis to measure how dependent or independent these factors are. Thus one can show, for example, that environmental problems, in a statistical sense, are very poorly correlated with security issues. One could naively draw the conclusion that environmental variables should not be relevant for decisions relating to security problems. How dangerous such conclusions might be in the context of nonlinear chaotic systems is illustrated in the following example: Assume we observe three time series (each one appropriately normalized) with apparently erratic time dependence. Visual inspection suggests that these time-series might not be independent (Fig. 2a). When we calculate the corresponding correlation between pairs of the three variables we find the results shown in Fig. 2b: Indeed, the x and y variables are highly correlated (up to 90%) whereas neither the x nor the y variable show any significant correlation with the third, z-variable. (Note the correlation is computed with a relative time-shift of τ.)

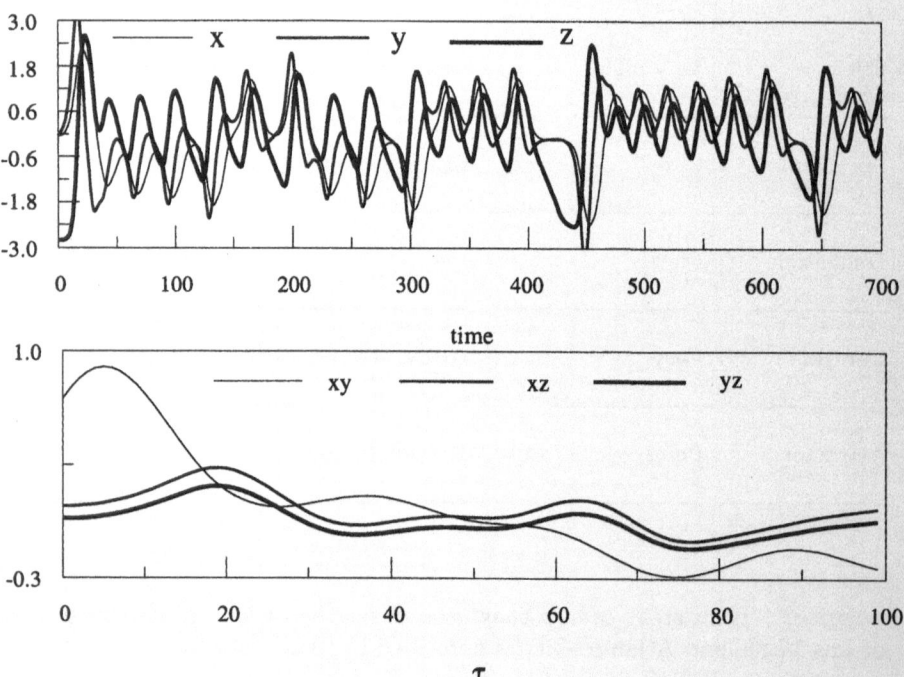

Fig. 2 (a) Times-series of the (normalized) x, y, z-coordinates of the standard Lorenz system. (b) Pair-correlation between each of the time-series of (a) with a relative time-shift.

Coming back to our initial example, let us now assume that the z-variable corresponds to a factor describing some environmental parameter. From the interpretation mentioned above we would expect that decisions which would have an effect on the z-parameter would not make much difference on the variables x or y.

4 Perturbation Analysis

In a computer experiment we now simulate a decision which would change at a given instant of time the value of the z-variable by an amount of less than a tenth of a percent. This impact is too small to be visible on the graph of Fig. 2. The results of this new simulation are shown in Figs. 3a,b: Figure 3a shows a superposition of the original z-variable (starting at $z_0 = (0.5, 0.5, 0.0)$) and the z-variable of the perturbed solution (starting at $z_0 = (0.5, 0.5, 0.01)$). As mentioned above, the time history for both solutions is basically the same for about 200 time steps (about 10 intrinsic cycles of the system which is a more characteristic time unit than the algorithm-dependent numerical time-step). Later on, the difference between the two curves increases until they have a basically independent evolution. In Fig. 3b we have the corresponding plot for the x variable, which is basically uncorrelated with the z-variable. Nevertheless we can see that after somewhat more than 200 time steps the two scenarios can have values which are completely (order unity) different from each other.

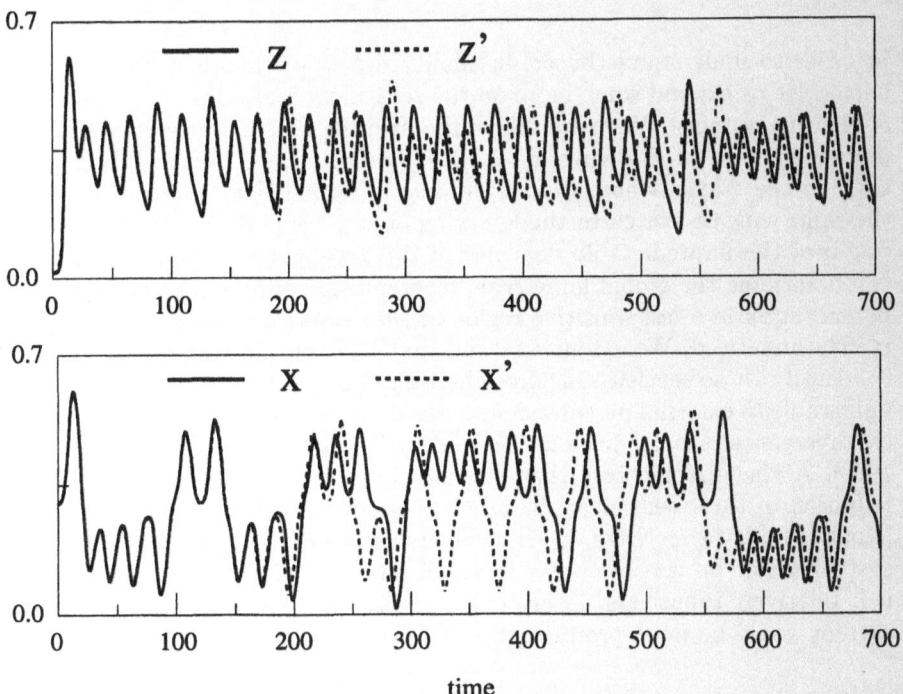

Fig. 3 (a) Superposition of the original z-variable and the z-variable of the perturbed solution z'. Initially the difference is only 0.01, i.e. less than 0.1%. (b) Superperposition of the original x-variable and the x-variable of the perturbed solution x'.

5 Strange Attractors

The explanation for this phenomenon is shown in Fig. 4: All three time-series are generated by the same set of equations, the well-known Lorenz equations [Lorenz, 1963].

Thus we can see that a geometrical analysis might prove much more powerful in cases like this than a quantitative statistical analysis. Thus we have to make sure that we capture all the relevant variables in a global model, even when there is no statistical correlation between some of them. This, of course, raises the question of nonlinear indicators for relevancy or redundancy, a question that is the topic of a very active area of current research. The conclusion from this example should not be that "everything is connected with everything" in a naive way. From the theory of synergetics [Haken, 1977, 1983] we know that there are enslaved variables and order parameters. For low-dimensional models we need to find out what these order parameters are and we need nonlinear methods to do that.

6 Sensitive Dependence on Initial Conditions

In order to understand the predictability aspect of chaotic systems a little better, let us expand some more on the sensitivity and divergence properties of chaotic systems: When we follow the orbit in Fig. 4, we can see that the motion on each of the wings of the Lorenz butterfly is fairly regular. Only close to the z-axis, where the system has to decide if it should continue on the right wing or switch to the left wing or vice versa do we directly see the origin of the unpredictable behavior of the Lorenz system: small influences can determine the global large scale future of the system. The same kind of perturbation in a less sensitive region of the Lorenz attractor could go completely unnoticed. We can understand that it is very important to find tools to identify those sensitive regions where the state of the system is extremely vulnerable to external perturbations. One of those nonlinear measures, the local divergence rate has been introduced by [Nicolis, Mayer-Kress and Haubs, 1983]. A visualization tool that provides a good global estimate of the distribution of these sensitivity parameters is given by the recurrence diagram [Koebbe and Mayer-Kress, 1991]. It allows us to anticipate which states of the system might be very sensitive to small fluctuations and in which domains it is relatively robust and insensitive and also where attempts to control the system would be most promising.

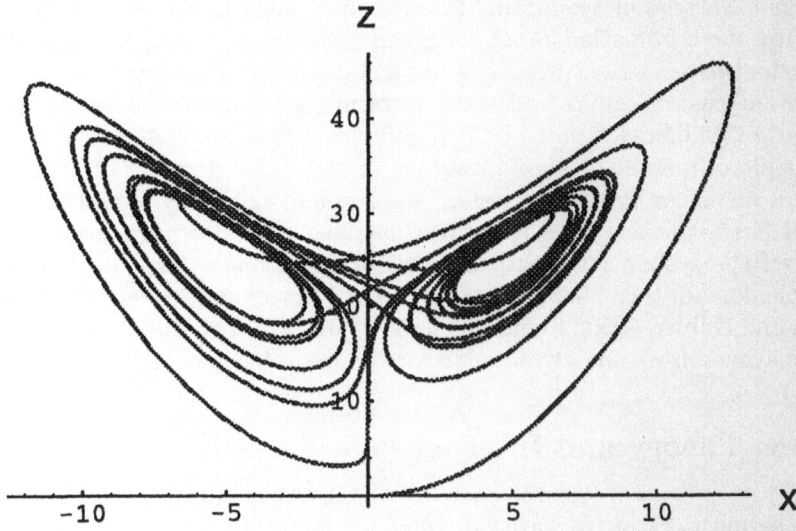

Fig. 4 $(x - z)$ projection of the Lorenz attractor.

7 Crises

The Lorenz system is very stable in the sense that we can always be confident that the small perturbations will only lead to solutions which are confined to the same global attractor. This can change, however, when we modify one of the internal systems parameters: we can now obtain configurations where the system would exhibit an attractor with a different type of sensitivity. Small changes now not only would determine the fate of individual orbits on the attractor, but the fate of the whole attractor can become sensitive to very small external perturbations [Mayer-Kress and Haken, 1984]. The system could be pushed into a new and qualitatively different attractor (internal crisis) or the system could collapse altogether in the sense that the solutions would diverge (external crisis) [Grebogi, Ott and Yorke, 1983].

8 Control of Chaos

A very interesting recent development in chaos theory is the control of chaos [Hübler, 1989; Hübler and Pines, 1992; Ott, Grebogi and Yorke, 1990]. Traditional methods of control theory could be shown to potentially lead to chaotic solutions as a result of the effort to control the system to achieve a stable state. A control force that has too large a feedback delay, could try to guide the system to a state that has changed due to the dynamics. The result is an overshooting that could lead to chaotic oscillations [Grossmann and Mayer-

Kress, 1989]. Instead of attempting to force the system to the desired state, one can use the information about the geometrical structure of the attractor of the system to drive the system close to a stable manifold of the goal state. Then the internal dynamics will assist in bringing the system closer to the desired state [Shinbrot et al., 1990]. A different approach that is based on the principle of matching takes advantage of the stable domains of chaotic systems: a nonlinear dynamical system will respond extremely sensitively to a control force which is close to its own intrinsic dynamics (nonlinear resonance). If the system is perturbed with such a resonant force while it is in a stable domain then this form of open loop control can be very effective [Jackson and Hübler, 1990]. Simulations show how extremely small forces can drive the system from one attractor to a more desirable one.

9   Chaos Theory and International Security?

The pioneering work of [Richardson, 1960] set the stage for subsequent attempts to analyze quantitatively many questions of strategic military and economic competition between – and among – nations. Some of his insights came from his work on a classic chaotic system, our weather. He had already anticipated many developments which were only realized decades later, when fast computers became available. Many phenomena that are exhibited by chaotic systems appear to have striking parallels in the interaction of human societies [Campbell and Mayer-Kress, 1991]. The interpretation that chaos always has a high risk of conflict is not always correct. For example in [Grossmann and Mayer-Kress, 1989] we discuss how low level bounded chaos can indicate a stabilizing, democratic discussion and interaction. Conditions that lead to an orderly but unbounded arms races can have a much higher risk of war. Some of these arguments have also been discussed in an model on the impact of SDI on the strategic arms race [Saperstein and Mayer-Kress, 1988]. This research indicated that very fast rates of deployment of SDI systems can lead to unstable and unpredictable solutions which can be interpreted as indications of chaotic processes. The numerical value of the predicted number of ICBMs for each side after a 30 year arms race could take any value between zero and 10000 missiles. More recently, the end of the cold war and its impact on the international security system have been studied. The end of the simple situation of a two-nation superpower competition leads to a multi-polar world in which alliances can form and change [Waltz, 1979]. Some elements of this problem have been studied by [Forrest and Mayer-Kress, 1991; Mayer-Kress, 1989] and will be discussed in the next section. Today the situation is much more complex, the potential futures of multiple, interacting, adaptive and intelligent agents will require new approaches, for example, for the international organizations [see e.g. the discussion in Mayer-Kress, Diehl and Arrow, 1995].

10 How Can Chaos Theory Be Applied to Crisis Management?

As we tried to explain above, chaos theory is intrinsically based on the treatment of problems as a nonlinear dynamical system. Solutions can be complex and typically they cannot be easily extrapolated from current trends. Qualitative changes in the response and, correspondingly, in the strategy will be more typical than quantitative competition in deterrence values measured in throw weight or mega-tonnage. National security interests have to be redefined and perhaps will have to be expressed differently than as budgetary numbers of large programs. Domestic economic interest might be increasingly in conflict with national security interests. Thus the notion of "preventive diplomacy" has to be taken seriously also by strategy planners. Many traditional areas of military power lose their role with respect to security issues whereas others increase their influence. A few years ago S. Kapitza gave a talk at Los Alamos National Laboratory where he made comparisons of the effect of nuclear weapons and national television. During the gulf war cable network news (CNN) played an important role in providing the global public with close to realtime information. Political and military decision-makers have to take this into account as a relevant factor for their decisions. It is also clear that decision makers will try to control that public information factor. This was relatively easy during the gulf war through classical means of censorship. This direct control of the news-media, however, can have nonlinear effects in that the public response to that control can change political parameters which can then act back onto the military decision makers. The complexity of this public information system will increase as multiple access to CNN type information becomes more available. Under those circumstances a plain censorship decision might not lead to the desired effect but could easily achieve the opposite outcome. Thus it would be very important in future crises to build careful models incorporating those factors which were absent or much less important in classical military planning.

The international drug industry represents another potentially important source conflict: The capital controlled by international drug organizations is of the same order of magnitude as some national defense budgets. Economic interests in the producing countries can often be in conflict with efficient drug policy enforcement measures. The current crisis in Peru, the most dominant producer of coca, could be studied under this perspective. While sophisticated technologies are developed to intercept ICBM missiles, it seems that modern technologies have only a 50% success rate of intercepting civilian ships and small airplanes. This fact might create a "window of vulnerability" which could be more dramatic than that in the context of ICBM surprise attacks. Another case where nonlinear model simulations could be helpful is in reevaluating the importance of specific international aid programs for national and international security, for example, a rational analysis of the amount that is spent for Israel as opposed to the amount spent to stabilize the situation in the former domain of influence of a collapsed superpower. If one superpower focuses on military strength without the corresponding economic backing, then in a system in which lobbies have traditionally had a large influence on policies there will be a very strong incentive for other powers to invest a large fraction of their defense budget for lobbying activities in that military power. For example, there seems to be some evidence of more or less direct influence of the Israeli and Japanese government on US politics. Therefore it seems feasible that superpowers in the classic style might not be sustainable any longer, largely as a result of economic and information realities. Maybe the gulf war, dominated by the US but largely financed by the allies, indicated a change in the global strategic arena.

11 Approaches from Nonlinear Dynamical Systems and Chaos Theory

How could we approach these problems from a nonlinear dynamical systems point of view? The first step certainly has to be a systems analysis. Much know-how in this sector has been developed in the context of systems dynamics, although there the analysis is guided by the restricted mathematical methods and certainly has to be improved and generalized. Guiding questions in that sense might be:

- "Who are the major actors in this crisis?"
- "How are these actors influencing each other?"
- "What are the external parameters that influence the decisions of the actors?"

Answering those questions and representing their answers in some structured form (graphical or symbolic) would constitute what we would call a conceptual model. Already at this conceptual level, recent progress in object oriented graphical user interfaces (GUIs) such as the Diagram tool described below, might provide assistance in a better understanding of the complexity of an evolving interacting system. Once this conceptual model has been established, often (but not always) it is straightforward to find quantitative variables to map this conceptual model onto a computational model. We can summarize: Decision makers will always have some model of the system of interest to assist reasoning during a crisis. Two (conflicting) factors contribute to the type of models chosen:

i) conceptual models use intuitive reasoning based on experience. They typically abstract the most relevant features from a sea of redundant information. They are very flexible to adapt to unforeseen changes and generalize easily to qualitatively new situations.

ii) Traditional analytical models try to anticipate as many factors as possible in as much detail as possible. Thereby they become very inflexible and practically useless in surprise situations. These models are typically not very adaptive.

Chaos theory suggests the use of models with local, short term predictability: Adaptability requires

- simple, low-dimensional models which should be intuitive
- fast and direct access to and integration of current global data
- multimedia user interface for efficient representation of the results
- global sensitivity analysis and identification of crisis domains
- efficient individual archiving and retrieval system. An efficient simulation environment would provide an object oriented integration of all of the elements described above
- conceptualization of complex developments without strict formalization
- access to global information systems that allow global estimation of parameters for low-dimensional chaotic models with global scan of scenarios
- links to detailed simulation systems where data and quantification requirements are satisfied.

In the remaining sections we want to give a brief description of some elements that might illustrate some of the points discussed above.

12 Adaptive Control and the Interaction of Two Agents:

The following example illustrates some of the issues under discussion [Hübler and Pines, 1993]:
Two agents interact in an environment which is influenced by three factors:

1   an unpredictable world, which accounts for all factors that are not directly accessible to modeling and control and provides a background unpredictable influence;
2   the actions of the first agent;
3   the actions of the second agent.

The goal of each of the two agents is to avoid crises and keep the environment predictable. Thus they have to adapt their strategies to their environment. Their success is measured by how well the future state of the environment can be predicted over some number of time steps.
The strategies that the agents have available are:

1   build a model of the environment with a level of complexity that the agents can choose. (The level of complexity is expressed as the number of parameters that go into the model);

2   the number of observations used to tune the model;
3   try to influence and control the environment to make it better predictable;
4   choose the type of order that the agent tries to impose onto the dynamics of the environment in order to achieve maximal predictability.

Since there are continuous unpredictable influences from the outside world and since the opponent also influences the environment in an unpredictable fashion, each agent has to continuously update the strategy. Simulations of this simple model provide some fairly interesting results: First of all both agents will try to control their environment/opponent but soon will realize that a "leader/follower" configuration allows better prediction for both agents. The attempt of a leader to introduce static order in the environment appears to be generally very unstable since it provides very little information about the internal state of the system and the opponent. Therefore the model of the dominating and controlling agent becomes increasingly worse until the mismatch between model ("ideology") and reality is large enough that the system becomes unstable and large scale fluctuations can build up. It appears that the most stable strategy of the leader is to impose a goal dynamics on the system which shows low level chaos. Thereby the controlling agent can continuously test an extended behavioral domain of the system and thereby keep the model up to date and close to reality (in the form of external world, immediate environment, and opponent.) From the perspective of the follower, fairly accurate short time predictions are also possible since the goal dynamics was assumed to be weakly chaotic. That means it is unpredictable enough to keep the follower alert but also structured enough to allow for successful adaptation and anticipation. When the system becomes too unpredictable or hopeless, a transition to a "no future" culture might be the response. It is very interesting to notice that the degree to which the external world is changing is essential for the degree of complexity of the most successful models for both leaders and followers: In a relatively stationary environment it pays to accumulate many data points in order to construct a model with a large number of parameters accurately. In the case of a rapidly changing world, however, sub-optimal, smaller models that need less input data and have fewer parameters to be estimated appear to become more successful, since they can be updated more rapidly, whereas a highly complex model can find itself making accurate predictions based on data or parameters that are already obsolete. Such conditions typically occur in crisis situations where simple adaptive models with tight links to global information systems might become superior in spite of their global character and lack of detail.

13 Global Information and Simulation Systems

The amount of information available about the state of our planet with all its subsystems is increasing dramatically. The data come both from direct observations as well as from computer simulations and more traditional methods. The representation and structuring of this rapidly changing information flood is a challenging and unsolved problem. From the theory of chaotic dynamics and the study of complex adaptive systems we have a sophisticated mathematical and computational tool-box available regarding global, invariant structures, bifurcations, analysis and response to external perturbations. These tools have been applied and tested for the analysis and modeling of a number of systems in a large variety of contexts. From a different angle they have been most successfully applied in *virtual realities* of educational computer games. Common to both of these examples is that they deal with a closed environment of theoretical or game pragmatic assumptions and parameters. What is lacking is some efficient interface to the real world of global dynamics. The technology for such an interface is currently being developed as global communication and information systems. High speed computer networks and information servers are commonly used as well as other areas of global networks. In this project we made a first attempt to utilize some of these modern communication, computer, and multi-media technologies to approach exactly that problem. A pilot project version of this approach[15] was presented as an interactive computer installation *Earth Station* at the 1991 ARS electronica [Mayer-Kress, 1991b]. Since this framework is very global and since our main interest is to demonstrate the power of modern computational tools with respect to adaptability and interactivity, we constantly updated the models, including some issues of current interest (through newspaper or through segments recorded from news broadcasts or received from the NetNews of the internet). At the same time we would drop issues which seemed to be no longer strongly coupled to the global dynamics, which, of course might prove completely wrong from some future perspective. In the following we want to describe several of the elements in our installation, which try to integrate generally available software and services into a useful information and simulation system. Our concept is mainly based on hierarchical network representations of current problem areas. Each node of the network corresponds to an object which can be of a very general type [Mayer-Kress, 1991a; 1992; 1993]:

- a network on a lower level
- images and charts, which themselves can act as background for new networks or be annotated with sound messages (news-clips) or general other programs

---

[15]Most of those features are very familiar today in the implementation of the Mosaic interface to WWW and tools for creating html files.

- simulation tools
- programs that connect to other units such as other computers, on which different types of programs can be launched, or general multi-media devices.

This framework is a natural background for geographic information systems, both static and dynamic. The links that connect these objects can be adapted very easily in a graphic, object-oriented programming style and thereby have a great advantage compared to classical world-model programs: The contents and the structure of the simulation and visualization tools can be easily updated as the state of the real world changes and evolves. Since our *Earth Station* pilot project in 1991, phase-transitions in the global information network have taken place and are still taking place. In a very qualitative sense one can notice a difference of the complexity and performance of the Internet and its information servers by asking a random question and then attempting to find the corresponding answer on the Internet. The perception in 1991 was that of a bad encyclopedia. In spite of the vast amount of information available, the chance of finding a specific piece of useful information was almost zero. Since then some percolation threshold has been crossed: the probability of finding the answer to a specific question is clearly not close to one yet, but it is definitely larger than zero. Nevertheless it is clear that the Internet will never become a database system of the type we are used to for example in a library. Information on the internet is intrinsically evolving, and not centrally controllable: any provider of information to the network can make a decision to make this information globally available, announce it, or change the read-permission such that it is only accessible to a specific group. This dynamic and complex feature of information on the global Internet has many aspects in common with how information is processed in biological brains.

14 The Global Brain Concept

The parallels between biological brains and the global Internet have been discussed in some details in [Mayer-Kress and Barczys, 1995]. In this section we want to summarize some elements and provide some recent examples and applications. [Russell, 1983] proposed a *Global Brain* that might emerge from a world wide network of humans who were highly connected through communications. He based his argument on the observation that throughout evolution qualitative transitions to a new level of organization have been observed to occur in several instances where a system attains approximately 10 billion (10^{10}) units that are tightly but flexibly coupled. Examples include the number of atoms in a bio-molecule, the number of molecules in a cell, and the number of cells in the cortex of the human brain. Since the world population (5.7 billion, 1994) is within an order of magnitude of 10 billion (10^{10})

and growing, the threshold for a new level of organization, by his arguments, could be reached soon[16]. Thus Russell saw the network of interconnected humans forming a Global Brain; we expand the concept to include computers — not only as communication links between humans, as Russell used them, but as active information-processors alongside humans. Simulations in this environment will have to include life-like, agent-based elements [see e.g. Langton et al., 1992].

14.1 Phase-Transition in the Information Flow on the Internet

In [Mayer-Kress and Barczys, 1995] we show maps of the information flow on subsystems of the global Internet with a strong concentration within the US. The main increase in the information flow rate has been observed in the past few years. In Fig. 5 we plot the NSFNET backbone traffic (see Fig. 4 in [Mayer-Kress and Barczys, 1995]) normalized by the number of Internet hosts for each of the information services WWW, Gopher, and WAIS[17].

The data are plotted on a logarithmic scale for the time since the end of 1992. We can observe that the hyper-media-based WWW grew at a rate of about 37% per month up to the first quarter of 1993. At that time an advanced user interface to WWW was introduced, NCSA's Mosaic. Since it is available for most of the common computer operating systems, it misses some of the advanced drag-and-drop features that we had available in 1991 for the NeXTstep-based Diagram interface. At the same time it created a very common, easy to use environment which appears to have triggered a sharp transition to a higher information flow rate which continued to grow at 25% per month (status: May 1994) [18]. During the transition period the temporary growth rates exceeded 600% per month. In the context of non-equilibrium phase transitions we would identify the number of Internet hosts as control parameter and the traffic per Internet host as order parameter [see Haken, 1977]. The analog of the free energy or generalized potential would be an *information deficit* that induces the generation of different unstable modes (in the sense of positive growth rates) of information flow or transport services (WWW, Gopher, WAIS). The interpretation of Fig. 5 could then be that there is a phase transition between two levels of information flow both of which have about the same stable growth rates with respect to the growth of the number of Internet hosts. The rapid transitions between those levels is induced by the release of X-Mosaic. The over-shooting during the

---

[16]At the beginning of 1994 the role of increase in the number of Internet users was much higher than the global population growth rate — 90% for Internet versus 1.7% for population. This difference in rates is so great that with a simple linear extrapolation, each human would also be an Internet user by the year 2001, see reference MIDS.

[17]The data are available at gopher://nic.merit.edu:7043/11/nsfnet/statistics.

[18]To our knowledge, Larry Smarr, NCSA, was the first to suggest that this might be a signature for a non-equilibrium phase transition [Smarr, 1994].

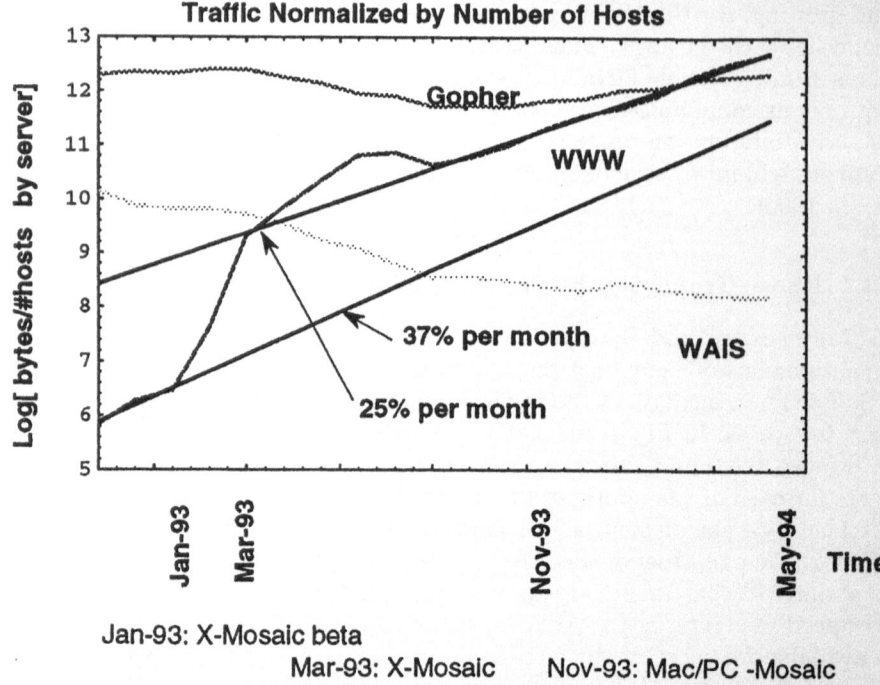

Fig. 5 NSFNET backbone traffic normalized by the number of Internet hosts for each service: WWW, Gopher, WAIS.

summer could either indicate a novelty/exploratory effect or it could be a seasonal effect, which could be tested during the following summers. Due to the highly connected nature of the Internet, information can propagate very rapidly through avalanche effects: Once a hyper-media document has been released and made known to an active group of users, a link to it will be included in the local hyper-media document of that group and consequently the news about the original document can spread rapidly like a global rumor. There are also key archive sites that are frequented by many users and thus have a dramatic effect on the visibility of specific documents. For example the document in [Mayer-Kress, 1994a] had an increase in access-frequency of more than 100% after it was announced in the NCSA "What's New?" page.[19]

14.2 Global Tele-conferencing on the Internet

In the context of biological brains it seems that perception of complex structures is achieved by the simultaneous rhythmic activation of a large cell assembly [see e.g. Barczys, 1993; Braitenberg and Schuez, 1991; Braun, Mayer-Kress and Miltner, 1994]. On a global scale one could argue that scientific

---

[19]URL: http://www.ncsa.uiuc.edu/SDG/Software/Mosaic/Docs/whats-new.html.

conferences play a similar role of establishing new insights into complex problems within the scientific community. On a different time-scale one could argue that this form of synchronization also happens through scientific publications. Hyper-media on the Internet provide a very important bridge between these established forms of scientific operation overcoming some of the shortcomings of both traditional methods (see Mayer-Kress, Bender and Bazik, 1994a, b). The role of the Internet will be especially important for the collaboration with researchers from developing countries, where handicaps both in terms of speedy availability of scientific journals as well as regular participation at international conferences can be severe. In a tele-conferencing experiment we created a Mosaic document [Mayer-Kress, Bender and Bazik, 1994a][20] that was transferred to the conference site prior to the presentation. The talk was given by telephone with visuals presented by a local operator displaying the Mosaic pages that were referred to by the speaker. This mixed mode of presentation provides a degree of real time interactivity at much lower communication demands than broadcast tele-conferencing together with a picture quality of the visuals that is much higher than that of slow scan TV. The World Hunger Program is now using this method regularly for its work.[21]

14.3 Future Extensions of the Global Brain Concept

14.3.1 Multiple Global Brains

In this last section we would like to add a few speculations about potential future developments of the Global Brain concept. It is natural to expect that many different, coexisting and perhaps competing global structures might evolve that would have different implementations of what we would assign to Global Brains.[22] The biological analog of such a development could be in the specialization into different brain areas fulfilling different functions.

14.3.2 Implication for Marketing

The notion of *Information Society* and *Information Industry* has been discussed in academic circles for quite a while. The visible presence of the global Internet provides some more practical views on these issues. One example might illustrate how an industry, that by many is considered to be a standard example of *low-tech*, namely agriculture, could be dramatically changed by the Global Brain paradigm. Currently we see considerable efforts, especially in Europe, to provide incentives for farmers to produce less quantity

---

[20]URL: http://www.ccsr.uiuc.edu/People/gmk/Papers/HungerConf/HungerConf.-html.

[21]See for example the WWW page of the Southern African Development Community (SADC) (http://netspace.org/hungerweb/SADC/).

[22]Peter Danielson was the first one who pointed this possibility out to us.

or produce in a way that complies with a number of ecological or environmental restrictions.[23] This immediately shows that the value of the product is now shifting from the physical nature of the product to the information about the product or its substitute. For example, the price that a farmer can achieve for NOT producing a quantity of grain could be higher than the market price for the product itself. That means that more valuable than the product itself is the information about it or its non-existence. This imposes questions about verification and control of compliance with the regulation which immediately smell like a rapidly expanding administrative and regulatory water-head. Other examples which might be more relevant are the information about how the product was produced. *Bread Labeling* is one example, where the choice made by the customer might depend on details of information about a loaf of bread that he or she is about to buy: what sort of chemicals have been used for the production? Under what conditions was it produced? Where was it produced and when?[24] But other factors might also be relevant for specific groups of customers, e.g. the political, religious, or other preferences of the farmer [Gell-Mann, 1991]. In the future we can expect that important goods such as food is individually labeled with an identification code similar to what is used today for overnight mail tracking. We can also anticipate that such a tracking could not be done centrally. In the case that every producer and dealer is connected to the global Internet, this enormous task can become much more feasible: Each farmer, say, can have a WWW/Mosaic document, in which all relevant information about the production and the labels of the delivered products are organized[25]. Consumer organizations or commercial vendors then can provide a service to the end-customer to convert individual preferences into an algorithm that will rank alternative choices for products. Instead of deciphering complicated product labels in miniature print, the individual customer would only have to use the identification number of the product and get an instantaneous evaluation with respect to the individual preference list[26]. The main effect on marketing, however, would be a reduction of need for regulation, since the almost real-time information that the producer makes available will combine effective self-regulation with successful marketing strategies. The information exchange on the Internet is effectively two-way: Every time a customer requests information about a certain product, the producer will obtain information about the nature of the electronic requests. By analysing access log-files one can even in quasi-real-time study, document and evaluate brows-

---

[23]In the US similar concepts are being discussed; for example, the present Conservation Reserve Program (CRP).

[24]Important especially in Eastern Europe or in cases of chemical spills.

[25]There are already many computerized book-keeping and activity report systems on the market, some are specifically developed for agricultural applications.

[26]If personal preferences are encoded in the credit-card number this system could substitute time-consuming and uninteresting checklists in restaurants and sandwich-bars about salad dressings and other trivial meal preferences.

ing behavior of potential customers and thereby try to anticipate and forecast trends in the market [see, e.g., Farmer and Sidorovich, 1988].

14.3.3 Extensions to Non-Human Species

A different direction that we might encounter in the evolution of the *Global Brain* is the inclusion of the intelligence of other than human species. If we look at the Internet map, we notice that the last large white areas (Africa and the former USSR) are rapidly filling in with Internet nodes. But the vast areas of the ocean are still disconnected from the Internet. There have been many reports on the intelligence of sea mammals and especially that of whales. There also have been many attempts at verbal or symbolic communication with dolphins and whales. The success was limited probably for a variety of reasons. We can imagine that the new hyper-documents with a point-and-click iconographic interface could perhaps provide a new channel for efficient communication. We suggested a computer interface for whales that is based on video-tracking: A monitor (or video screen) displays a superimposed image of a computer window and the video image of the whale. A specific feature in the video image[27] can be tracked using commercially available software such as "Mandala" of the Vivid Group that has been used in virtual reality environments for years. The whale gets optical feedback about the connections between the motion of the tracked feature and the motion of the cursor on the screen. Collisions between the cursor and icons on the screen can be detected and used to trigger specific events. In a computer interface context this would correspond to a (video)-mouse without any buttons. The button-down event could be triggered by a specific (learned) vocalization of the whale[28] simultaneously with the collision between cursor and the icon. A main advantage of such an interface would be that it is independent of direct exposure to physical interaction with the whales or water. We can think of solar powered internet stations floating on the ocean with (Iridium) satellite hook-up and (LCD) screens with video/microphone interface.[29] With such an interface a common language is available that could be used for joint global tasks. For example surveys related to global change problems could possibly be done in cooperation with whales. Observation of whale behavior already provides indirect information about ecological changes in the ocean:

(...) In peak years during the late '80s, more than 200 humpbacks would be identified around Stellwagen Bank in Massachusetts Bay in a season. Last

---

[27]In the simplest case and in a controlled environment this could be a ball of a fluorescent color attached to a bite-plate [Scarpuzzi, 1994] that the whale can move around.

[28]See, e.g., [Kinsey, 1991] and http://www.ccsr.uiuc.edu/People/gmk/Projects/Or-cas-SW/

[29]Because of the purely visual interface no mechanical contact or exposure to corroding seawater would be necessary.

year, just 69 were recorded, and this summer humpbacks are an even rarer sight. At the same time, more and more are beingseen in the waters around Jeffreys Ledge off the coast of New Hampshire and Maine. The movement of the humpbacks is drawing much scientific interest, offering clues not only to how the endangered marine mammals survive, mate, and behave, but also to changes that may be taking place in the seas — the overfishing of species the whales eat, or in water temperature or circulation. In the murky and poorly understood deep-sea world, whales are one of the few visible barometers of change. "Whenever a shift like this happens, it means something's happening in the ecosystem," says Mason Weinrich, a zoologist who heads the Gloucester-based Cetacean Research Institute and has been studying humpbacks for a decade. (...)
[Jeffreys Ledge, The Boston Globe, Science & Technology, 25, 08/15/94]

But eventually one can imagine a world, where the global cyber-space is shared by all intelligent beings of this planet.

References

Barczys, C. (1993): Spatio-Temporal Dynamics of Human EEG During Somatosensory Perception. Ph.D. Dissertation, University of California at Berkeley.

Bass, T. (1985): The Eudemonic Pie. Science of mind publishers, Los Angeles.

Berge, P., Pomeau, Y., Vidal, C. (1984): L'Ordre dans le Chaos. Hermann, Paris. (English translation: Wiley, 1986.)

Braitenberg V., Schuez, A. (1991): Anatomy of the cortex : statistics and geometry. Springer-Verlag. Berlin, New York.

Braun, C., Mayer-Kress, G., Miltner, W. (1996): Wavelet based measures of short-term coherence in 40 Hz oscillations of human EEG during associative conditioning. To be published.

Campbell, D., Mayer-Kress, G. (1991): Chaos and Politics: Simulations of Nonlinear Dynamical Models of International Arms Races. Proceedings of the United Nations University Symposium *The Impact of Chaos on Science and Society.* Tokyo, Japan.

Farmer J.D., Sidorovich, J. (1988): Exploiting chaos to predict the future. In *Evolution, Learning, and Cognition*, ed. Y.C. Lee. World Scientific Press, 277.

Forrest, S., Mayer-Kress G. (1991): Using Genetic Algorithms in Nonlinear Dynamical Systems and International Security Models. In: *The Genetic Algorithms Handbook*, L. Davis, (Ed.). Van Nostrand Reinhold, New York.

Gell-Mann, M. (1991): Private conversation, Santa Fe Institute.

Grebogi, C., Ott, E., Yorke, J. (1983): Crises, Sudden Changes in Chaotic Attractors and Transient Chaos. Physica D **7**, 181.

Grossmann, S., Mayer-Kress, G. (1989): The Role of Chaotic Dynamics in Arms Race Models. Nature **337**, 701—704.

Haken, H. (1977): Synergetics: An Introduction. Springer, Berlin.

Haken, H. (1983): Advanced Synergetics. Springer, Berlin.

Hübler, A. (1989): Adaptive Control of Chaotic Systems. Helv. Phys. Acta **62**, 343.

Hübler, A., Pines, D. (1993): Prediction and Adaptation in an Evolving Chaotic Environment, eds. G. Cowan, D. Pines, and G. Meltzer. Addison-Wesley, Reading, MA.

Hübler, A. (1992): Modeling and Control of Complex Systems: Paradigms and Applications. In *Modeling Complex Phenomena*, ed. by L. Lam. Springer, New York.

Jackson, E.A., Hübler, A. (1990): Periodic Entrainment of Chaotic Logistic Map Dynamic. Physica D **44**, 407.

Kinsey, M.M. (1991): Evidence of vocal sequencing by killer whales, Orcinus Orca, MS thesis, Univ. San Diego.

Koebbe, M., Mayer-Kress, G. (1991): Use of Recurrence Plots in the Analysis of Time-Series Data in Nonlinear Modelling and Forecasting. *SFI Studies in the Sciences of Complexity*. Proc. Vol. XII, eds. M. Casdagli, S. Eubank. Addison-Wesley, Reading, MA.

Langton, C.G., Taylor, C., Farmer, J.D., Rasmussen, S., eds. (1992): Artificial Life II. Addison Wesley, Reading, MA.

Lorenz, E. (1963): Deterministic nonperiodic flow, J. Atmos. Sci. **20**, 130–141.

Marti, J. (1991): Chaos might be the new world order. em Utne Reader **n48**(30).

MIDS: Matrix Information and Directory Services, mids@tic.com, Austin, TX. [The figure is available as *33_in_year_2001.gif* from: gopher://ietf.cnri.reston.va.us/11-/isoc.and.ietf/charts/metrics-gifs]

Mayer-Kress, G., Haken, H. (1984): Attractors of Convex Maps with Positive Schwarzian Derivative in Presence of Noise, Physica D **10**, 329–339.

Mayer-Kress, G. (1989): A Nonlinear Dynamical Systems Approach to International Security. In: *The Ubiquity of Chaos* (S. Krasner, ed.). Proceedings AAAS conference. San Francisco.

Mayer-Kress, G. (1991a): Nonlinear Dynamics and Chaos in Arms Race Models. Proc. Third Woodward Conference: *Modelling Complex Systems* (L. Lam, ed.). San Jose.

Mayer-Kress, G. (1991b): EarthStation. In: *Out of Control*, Ars Electronica (K. Gerbel, ed.), Landesverlag Linz, Linz.

Mayer-Kress, G., Barczys, C. (1995): The Global Brain as an Emergent Structure from the Worldwide Computing Network, and its Implications for Modelling. The Information Society, **11**(1).

Mayer-Kress, G. (1992): Chaos and Crises in International Systems. Technical Report CCSR-92-15. Proc. of *SHAPE Technology Symposium on Crisis Management*. Mons, Belgium.

Mayer-Kress, G. (1993): Global Information Systems and Nonlinear Methods in Crisis Management. In: *1992 Lectures in Complex Systems* (L. Nadel and D.L. Stein, eds.). Santa Fe Institute Studies in the Sciences of Complexity, Lecture Volume V. Addison-Wesley, Reading, MA, 531–552.

Mayer-Kress, G., Diehl, P., Arrow, H. (1995): The United Nations and Conflict Management in a Complex World. *http://www.ccsr.uiuc.edu/People/gmk-/Papers/UNCMCW/*, (in progress).

Mayer-Kress, G., Bender, B., Bazik, J. (1994a): Hyper-Media on the Internet as a Tool for Approaching Global Problems: A Tele-Conferencing Experiment. Code No 94-0023 *Unpublished Scholarly Papers, Information and Communication, CIESIN Human Dimensions Kiosk, http://WWW.ciesin.org/kiosk/home.html.* [Presentation given via the Internet and telephone at the Hunger Research Briefing and Exchange, Brown University, April, 1994]

Mayer-Kress, G., Bender, W., Bazik, J. (1994b): A Tele-Conferencing Experiment with WWW/Mosaic. Technical Report CCSR-94-25. [*Proc. 2nd International World Wide Web Conference*. Chicago (see http://www.ncsa.uiuc.edu/SDG-/IT94/IT94Info.html).]

Mayer-Kress, G., Choi, I., Bargar, R. (1993): Sound Synthesis and Music Composition using Chua's Oscillator. Proc. NOLTA93, Hawaii.

Mayer-Kress, G., Choi, I., Weber, N., Bargar, R., Hübler, A. (1993): Musical Signals from Chua's Circuit. IEEE Transactions on Circuits and Systems **40**, special issue on *Chaos in Nonlinear Electric Circuits*, 688–695.

Mayer-Kress, G., Bargar, R., Choi, I. (1992): Musical Structures in Data From Chaotic Attractors. Technical Report CCSR-92-14. *Proceedings of the International Symposium on the Auditory Display (ICAD92)*. Santa Fe, NM. Proceedings Volume XVIII Santa Fe Institute Series in the Sciences of Complexity, Addison Wesley, Reading, 1994.

Mayer-Kress, G. (1994a): Chua's Oscillator: Applications of Chaos to Sound and Music. *URL: http://www.ccsr.uiuc.edu/People/gmk/Projects/ChuaSoundMusic.*

Mayer-Kress, G. (1994b): "Non-Equilibrium Phase Transitions of Information Flow on the Internet?". *URL: http://www.ccsr.uiuc.edu/People/gmk/Projects/Web-Stats/webstats.html.*

Nicolis, J., Mayer-Kress, G., Haubs, G. (1983): Non-Uniform Chaotic Dynamics with Implications to Information Processing. Z. Naturforsch. **38a**, 1157–1169.

Ott, E., Grebogi, C., Yorke, J. (1990): Controlling chaos. Phys. Rev. Lett. **64**(11), 1196–1199.

Richardson, L.F. (1960): Arms and insecurity. Boxwood, Pittsburgh.

Russell, P. (1983): The Global Brain: speculations on the evolutionary leap to planetary consciousness. Houghton Mifflin, Boston, MA.

Saperstein, A., Mayer-Kress, G. (1988): A nonlinear dynamical model of the impact of SDI on the arms race. J. Conflict Resolution **32**, 636–670.

Schatz, B., Hardin, J. (1994): NCSA Mosaic and the World Wide Web: Global Hypermedia Protocols for the Internet. Science **265**, 895–901.

Scarpuzzi, M., Al Garver (1994): Private discussions, SeaWorld, San Diego.

Shinbrot, T., Ott, E., Grebogi, C., Yorke, J. (1990): Using Chaos to Direct Trajectories to Targets. Phys. Rev. Let. **65**(26), 3215—3218.

Smarr, L. (1994): Private communication, Urbana.

Waltz, K. (1979): Theory of International Politics. Addison-Wesley, Reading, MA.

Index